SOIL REINFORCEMENT FOR ANCHOR PLATES AND UPLIFT RESPONSE

SOIL REINFORCEMENT FOR ANCHOR PLATES AND UPLIFT RESPONSE

HAMED NIROUMAND

Butterworth-Heinemann is an imprint of Elsevier
The Boulevard, Langford Lane, Kidlington, Oxford OX5 1GB, United Kingdom
50 Hampshire Street, 5th Floor, Cambridge, MA 02139, United States

Copyright © 2017 Elsevier Inc. All rights reserved.

No part of this publication may be reproduced or transmitted in any form or by any means, electronic or mechanical, including photocopying, recording, or any information storage and retrieval system, without permission in writing from the publisher. Details on how to seek permission, further information about the Publisher's permissions policies and our arrangements with organizations such as the Copyright Clearance Center and the Copyright Licensing Agency, can be found at our website: www.elsevier.com/permissions.

This book and the individual contributions contained in it are protected under copyright by the Publisher (other than as may be noted herein).

Notices
Knowledge and best practice in this field are constantly changing. As new research and experience broaden our understanding, changes in research methods, professional practices, or medical treatment may become necessary.

Practitioners and researchers must always rely on their own experience and knowledge in evaluating and using any information, methods, compounds, or experiments described herein. In using such information or methods they should be mindful of their own safety and the safety of others, including parties for whom they have a professional responsibility.

To the fullest extent of the law, neither the Publisher nor the authors, contributors, or editors, assume any liability for any injury and/or damage to persons or property as a matter of products liability, negligence or otherwise, or from any use or operation of any methods, products, instructions, or ideas contained in the material herein.

British Library Cataloguing-in-Publication Data
A catalogue record for this book is available from the British Library

Library of Congress Cataloging-in-Publication Data
A catalog record for this book is available from the Library of Congress

ISBN: 978-0-12-809558-4

For Information on all Butterworth-Heinemann publications
visit our website at https://www.elsevier.com/books-and-journals

Publisher: Joe Hayton
Acquisition Editor: Andre Gerhard Wolff
Editorial Project Manager: Jennifer Pierce
Production Project Manager: Kiruthika Govindaraju
Cover Designer: Mark Rogers

Typeset by MPS Limited, Chennai, India

CONTENTS

BIOGRAPHY

HAMED NIROUMAND

Dr Hamed Niroumand is an Assistant Professor at Department of Civil Engineering, Buein Zahra Technical University. He is currently the Vice-Chancellor for the Research and Academic, Buein Zahra Technical University. His main fields of research are geotechnical engineering, earth anchors, deep foundation, numerical analysis, sustainable development, and nano-materials. He is a project manager and professional engineer in various geotechnical and earth buildings projects. In the years 2011, 2012, 2013 and 2015, he received various awards including four medals and international awards for his inventions and researches and first position in the research section at the National Iranian Young Inventor and Researcher Festival 2012 and the first in the research section at the National Iranian Youth Festival 2012 and 2013. He received the best researcher award in the Ministry of Road and Urban Development (MRUD) in 2016. He was the Chairman and Head Director of international/national conferences of civil engineering in almost 20 cases that were held in various countries. He has chaired sessions in several international/national conferences and festivals in various countries and has presented various research papers in many conferences around the world. He has published many papers in journals and conferences. He is in the editorial team and a reviewer in various scientific journals. He has around 15 inventions that are currently patented/patent pending. He has also received many awards for his researches.

PREFACE

Anchor plates are typically fixed to the structure and embedded in the ground to effective depth so that they can resist various load. This book is focused to improve the existing anchor plate system by introducing geosynthetics. The anchor plates consist of three different shapes that represent the square, circular, and the rectangular. This book described the role of some geosynthetics

The book contains 6 chapters. The 1st chapter describes the all types of earth anchors in geotechnical applications. The all methods and techniques of anchor plates describe in Chapter 2, Literature Review. Chapters 3, Research Methodology, show the various methodology steps of anchor plates in reinforced soils. Chapters 4, Experimental and Numerical Results, present some useful information regarding uplift performance of anchor plates in reinforced cohesion less soils. Chapter 5, Comparison Between Existing Theories and Experimental Works, and Chapter 6, Conclusions and Recommendations describe the main achieves of anchor plates in reinforced cohesion less soils.

The authors of this book hope that they success in providing readers with useful information regarding anchor plates of different types.

Hamed Niroumand

ACKNOWLEDGMENT

Many friends and colleagues contributed to this book. A special note of thanks goes to my friends, colleagues, Khairul Anuar Kassim and Mr Hadi Arabi who helped with the picture design. We welcome any positive suggestions from those who read this book.

CHAPTER 1

Introduction

1.1 INTRODUCTION

Many structures experience overturning moments due to lateral loads, which result in a combination of tension and compression responses at the foundation level. The designs of some structures need various systems to resist uplift forces. Under such conditions, effective and safe design methods can be achieved through the use of tension elements. These elements are referred to as ground anchors. The elements are typically fixed to the structure and embedded in the ground to effective depths, so that they can resist uplifting loads. Soil anchors are typically used to resist such uplift loads, although they are also useful as measures to increase soil stabilization. These systems are used for retaining walls (Fig. 1.1), transmission towers (Fig. 1.2), foundations (Fig. 1.3), sea walls (Fig. 1.4), and pipelines (Fig. 1.5). The soil anchors involved are of different types such as screw anchors (Fig. 1.6), grout-injected anchors (Fig. 1.7), anchor plates (Fig. 1.8), anchor piles (Fig. 1.9), and irregular-shaped anchors (Fig. 1.10).

The above examples indicate that few soil anchors are used to transfer loads from superstructures to denser soils; this is because the presence of lateral loads would induce an uplift reaction on the soil anchors. The design requirement is therefore based on both compressive and tensile criteria for the successful implementation of a structure's response, although tensile criteria are more important compared to compressive criteria in soil anchors. To fulfill these criteria, symmetrical anchor plates are usually employed. They are more effective compared to other soil anchor types. Anchor plates can be cast-in-place by excavation. Construction of symmetrical anchor plates in cohesionless soil is also comparatively easier than in cohesive soils. With the increasing use of cast-in-place anchor plates to resist uplift forces, the need for rational design procedures becomes apparent; these would account for soil properties at the intended anchor plate location and the anchor-plate – soil predicted response.

Soil Reinforcement for Anchor Plates and Uplift Response.
DOI: http://dx.doi.org/10.1016/B978-0-12-809558-4.00001-7
© 2017 Elsevier Inc.
All rights reserved.

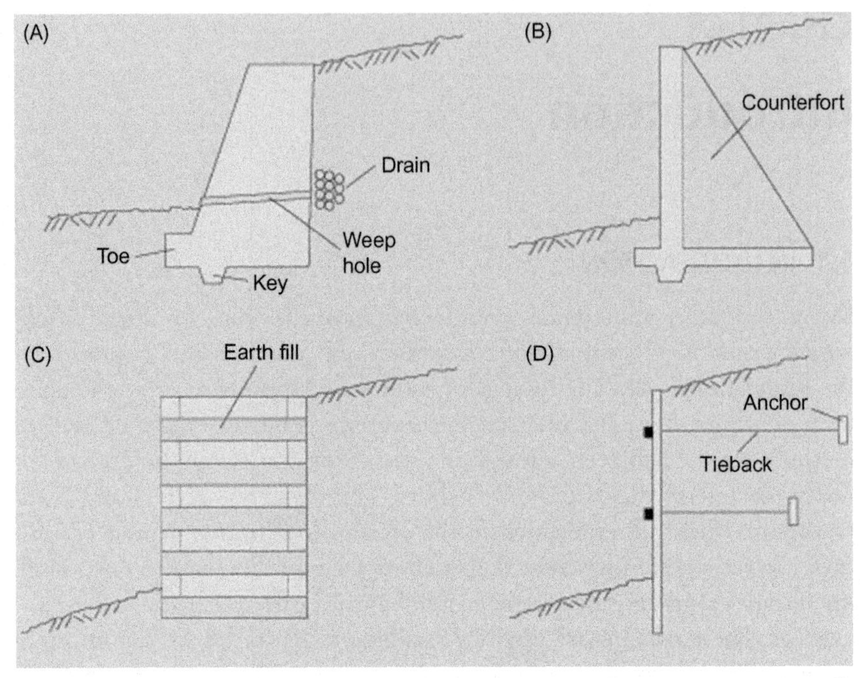

Figure 1.1 Symmetrical anchor plates in retaining walls. (A) Retaining wall; (B) Counterfort; (C) Earth Fill; (D) Tied-Retaining Wall.

Figure 1.2 Transmission towers subjected to uplift forces.

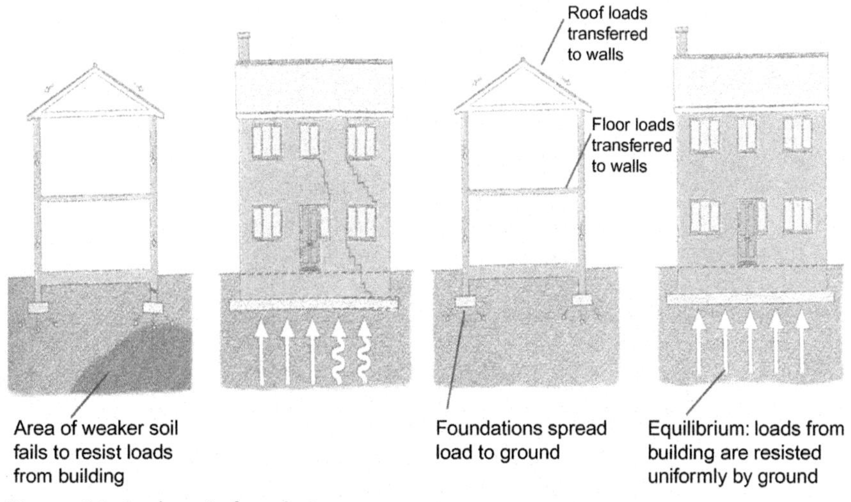

Figure 1.3 Anchors in foundations.

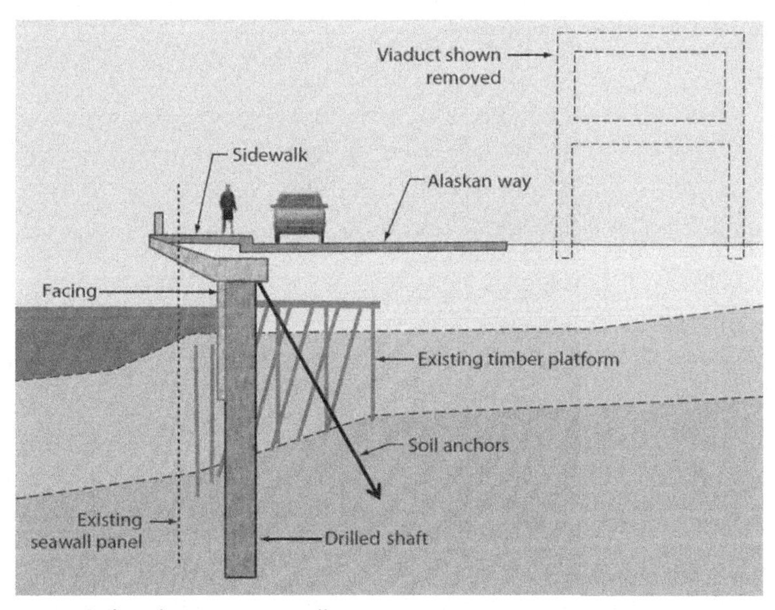

Figure 1.4 Soil anchors in a sea wall.

Most of the soil anchoring in various countries is grout based. The installation can take several days because grout needs special conditions in non-reinforced soils in construction. The symmetrical anchor plate can be installed in soils without having to grout it. Based on existing studies,

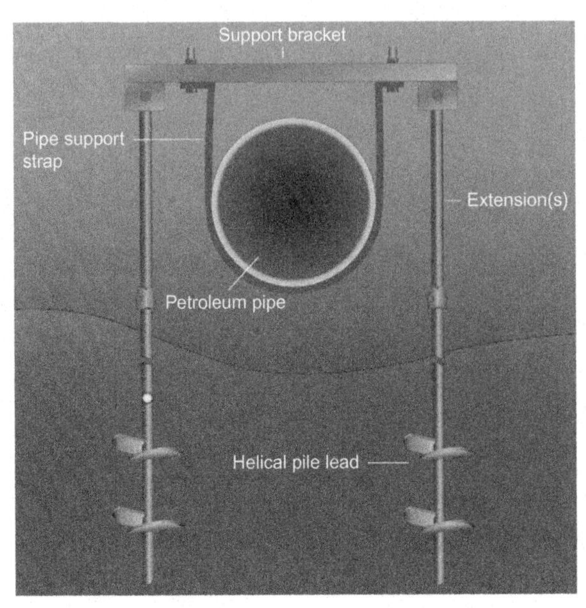

Figure 1.5 Soil anchors in a pipeline.

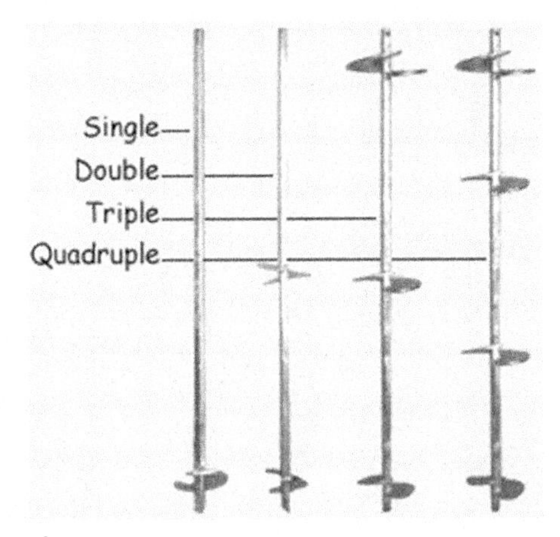

Figure 1.6 Screw anchors.

most ground-anchoring systems are concerned with the uplift response and failure zone in cohesionless soil. The use of geosynthetics and grid-fixed reinforcement (GFR) is expected to enhance the performance of symmetrical anchor plates by improving the uplift response and failure

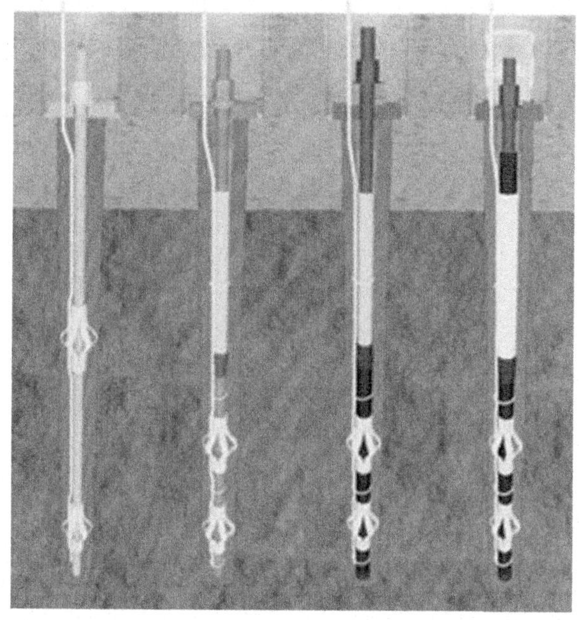

Figure 1.7 Grouted anchors.

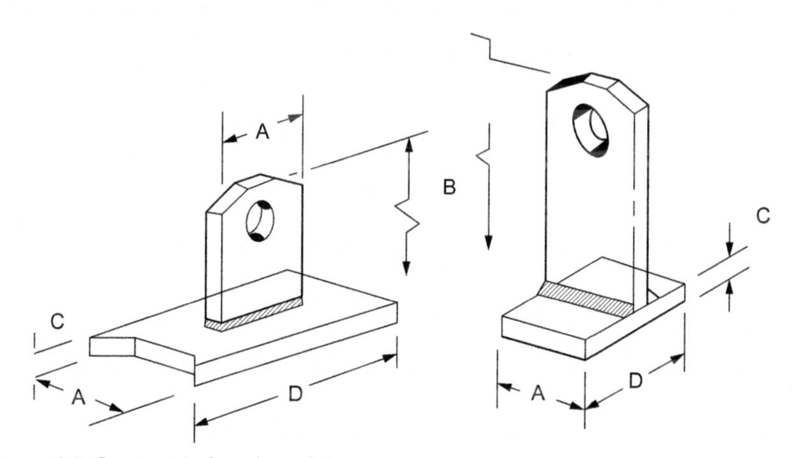

Figure 1.8 Symmetrical anchor plates.

zones. The pullout resistance of reinforcing materials is one of the most significant elements in increasing the uplift response of geogrid soils. In this book, a new reinforcing system that involves tied-up elements joined to the geogrid to increase the uplift resistance of geosynthetics is

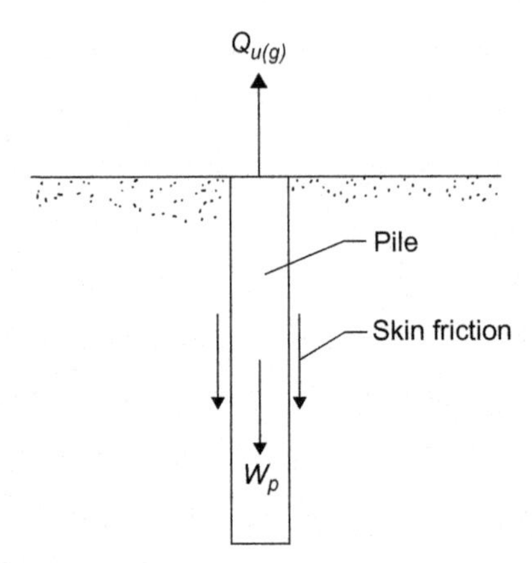

Figure 1.9 Anchor piles.

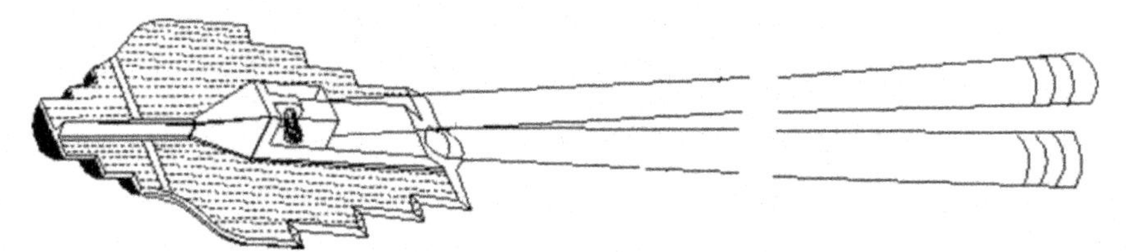

Figure 1.10 Irregular-shaped anchor.

described. GFR is a new tied-up system with innovative design, methods, and materials, using fiber-reinforced polymer (FRP) which can tie the geosynthetics into soils. It is expected that using geosynthetics, GFR, and numerical analysis will increase the uplift response, cost-effectiveness, and speed of analysis and design in construction projects.

CHAPTER 2

Literature Review

2.1 INTRODUCTION

A review of previous theoretical and experimental work and the analysis of an anchor plate embedded in sand experiencing uplift loading are considered in this chapter. Researchers such as Mors (1959), Giffels, Graham, and Mook (1960), Balla (1961), Turner (1962), Ireland (1963), Sutherland (1965), Mariupolskii (1965), Kananyan (1966), Baker and Konder (1966), Adams and Hayes (1967), Andreadis, Harvey, and Burley (1981), Dickin (1988), Frydman and Shaham (1989), Ramesh Babu (1998), Krishna (2000), Fargic and Marovic (2003), Merifield and Sloan (2006), Dickin and Laman (2007), Kumar and Bhoi (2008), Kuzer and Kumar (2009) were concerned with the general solution especially for an ultimate uplift capacity in sand.

One of the applications of soil anchors was in transmission towers. This application was a preliminary for much of the initial research into anchor behavior (Balla, 1961). This simple design came at considerable cost and, as a result, research was undertaken in order to find a more inexpensive design solution. As the range of anchor applications has broadened to include different support structures, a more concerted research effort has evolved to provide an inexpensive and competitive alternative to these foundations. Generally, research into the behavior of anchors can be either in the form of experimental or numerical/theoretical studies. In this chapter no attempt is made to present complete information on all research in soil anchors, rather a more selective summary of related research is presented. However, the theory for the capacity of grouted anchors is also applicable to the typical soil pullout behavior for symmetrical anchor plates. It will become clear that past research has been experimentally based and, as a result, current design practices are largely intuitive. In contrast, very few thorough numerical analyses have been performed to determine the ultimate uplift loads of different systems. However, of the numerical studies that have been published in previous literatures, few can be considered to be accurate.

Soil Reinforcement for Anchor Plates and Uplift Response.
DOI: http://dx.doi.org/10.1016/B978-0-12-809558-4.00002-9
© 2017 Elsevier Inc.
All rights reserved.

2.2 REVIEW OF PREVIOUS EXPERIMENTAL WORKS

This section presents the history of the performance of anchor plates in sand in many experimental works. It is based on different previous researches, from the earliest to the most recent. The main role of this most recent experimental research presented below, is focused on the prediction of the anchor's behavior and the force in the sand. Although there are not entirely sufficient substitutes for a full-scale field testing, tests performed in laboratories have the advantage of allowing close control of at least some of the parameters encountered in the research. This section discusses the field studies regarding the horizontal anchor plates, reported by Balla (1961), Hanna and Carr (1971), Das and Seeley (1975), Andreadis et al. (1981), Ovsen (1981), Murray and Geddes (1987, 1989), Fryman and Shamam (1989), Dickin (1988, 1994), Tagaya, Scott, and Aboshi (1988), Sarac (1989), Bouazza and Finlay (1990), Sakai and Tanaka (1998), Pearce (2000), Fargic and Marovic (2003), Dickin and Laman (2007), Kumar and Bhoi (2008). The results derived from the tests performed in the laboratory are typically a specific problem and they are difficult to extend and develop to field problems because of the different materials or the geometric parameters used in the field scale.

In order to research the performance of the ultimate uplift load capacity of the horizontal anchor plate, many numerical analysis and laboratory works have been developed. However, there are relatively few papers in the technical literature which briefly deal with horizontal anchor plates in sand. For numerical analysis purposes, the limited ultimate capacity of the horizontal anchor plate embedded in the sand can be obtained due to certain numerical and experimental researches. Many laboratory tests were performed on small as well on large models, in order to find an appropriate solution. During the last 50 years, various researchers have conducted several laboratory studies to provide a better understanding and to predict the ultimate uplift load of anchors in a range of soil types.

Most of the previous experimental research focuses on predicting the behavior of anchors and their capacity in sand. Even though the substitutes for full-scale field testing are not sufficient, tests performed on a laboratory scale have the advantage of allowing close control of at least some of the parameters encountered in the research. In this way, in some cases the behavior observed in the laboratory can be valuable and helpful in developing a good understanding of the performance at larger scales. Furthermore, observations made in the laboratory can be used in conjunction with numerical analysis to develop different theories. These solutions can then be applied to solve a wide range of problems. Experimental investigations on the plate anchor's behavior have typically adopted one of the two following selections: conventional methods under normal gravity conditions or centrifuge systems. Centrifuge systems use the laws of physical scale to match the model and the prototype behavior. Thus, a particular anchor size buried at a sufficient depth can also be used to investigate a range of burial depths, simply by verifying the stress field. As for the rest, the setup is subjected to a static gravitational force that equals 1.00 g. By keying the model in a centrifuge motion, researchers obtain gravitational forces greater than 1.00 g, which achieves the required stress field and, in turn, they simulate the situation stresses for different burial depths. Unfortunately, due to the basic equipment and the setup costs, only a few institutions possess this kind of tool, in use and in activation. Unfortunately, full-scale testing of the foundations is very expensive, consumes a lot of time, and in most cases it is difficult and only provides approximate results. For these reasons, testing is typically limited to small-scale model tests as these provide an economical and convenient alternative. The above-mentioned methods are the most common for sand, and also affect the centrifuge testing. Both methods have their own advantages and disadvantages, and these must be always taken into consideration, when interpreting the results from experimental studies of anchor behavior. The following sections provide a brief abstract of early and recent experimental researches regarding the behavior of the plate anchors in different types of soils.

A failure mechanism was adopted to determine the uplift load, by considering the equilibrium of the soil mass above the horizontal anchor plate and the shape of the failure in the adopted failure surface. Based on the underlying adoption, these methods of analysis can be separated into two subitems:

1. The soil cone method (Mors, 1959) in which the failure surface consists of a small cone extending from the anchor edges up to the intersection of the soil surface at an angle of $\left(45 - \frac{\varnothing}{2}\right)$ degrees, as shown in Fig. 2.1.

   The anchor force is by default equal to the weight of the soil, contained within the area of the assumed failure surface. Any shearing resistance and the force developed along the failure surface are very small.

2. The friction cylinder method (Downs & Chieurzzi, 1966), in which the failure is assumed to occur along the surface of a cylinder of soil

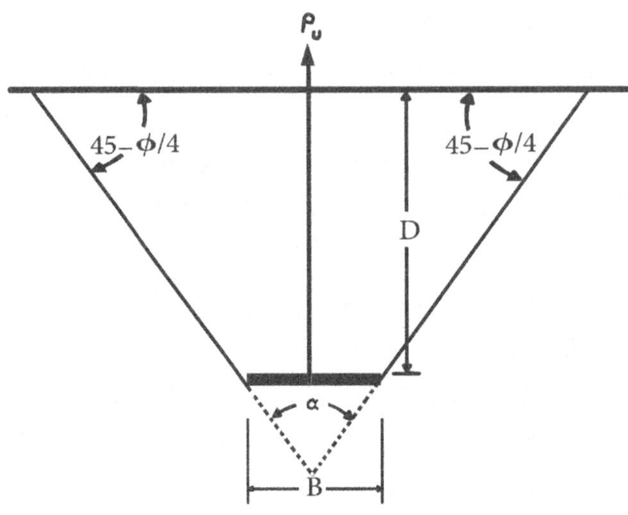

Figure 2.1 Failure surface assumed. *Source: Mors, H. (1959). The behaviour of most foundations subjected to tensile forces. Bautechnik, 36(10), 367–378.*

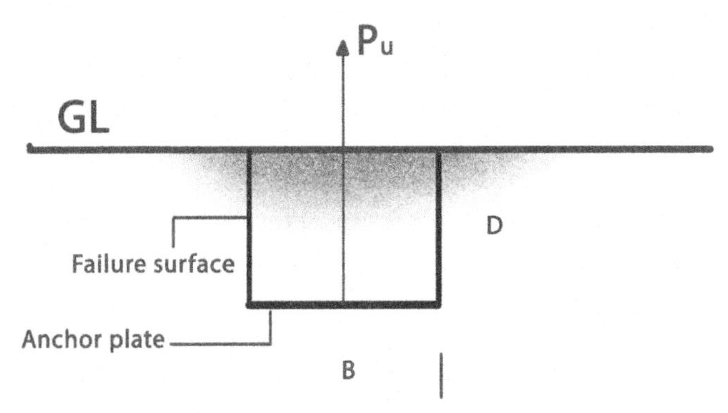

Figure 2.2 Failure surface assumed. *Source: Downs, D. I., & Chieurzzi, R. (1966). Transmission tower foundations. Journal of the Power Division, 88(2), 91–114.*

above the anchor. The anchor capacity is supposed to equal the sum of the weight of the soil, being contained within the area of the default cylinder to the failure surface, the frictional resistance, and the force derived along the failure surface, as shown in Fig. 2.2.

Subsequent variations on these early theories have been proposed, such as the proposal of Balla (1961), who determined the shape of the slip surfaces for the horizontal shallow anchors in dense sand. He proposed a numerical method for estimating the force of the anchors based on the observed shapes of the slip surfaces, as illustrated in Fig. 2.3.

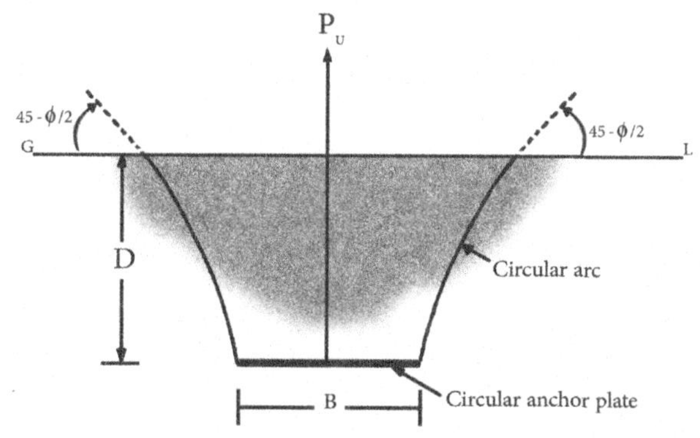

Figure 2.3 Failure surface assumed. *Source: Balla, A. (1961). The resistance of breaking-out of mushroom foundations for pylons. Proceedings of the Fifth International Conference on Soil Mechanics and Foundation Engineering, 1, 569–576.*

Baker and Kondner (1966) approved Balla's law findings regarding the behavioral difference between the deep and shallow anchors in dense sand.

Sutherland (1965) showed results for the pullout capacity of the 150-mm horizontal anchors in loose and dense sand, as well as large-diameter shafts in medium-dense and dense sand. It was concluded that the mode of failure differed according to the sand density and Balla's analytical approach may give reasonable results only in sand characterized by intermediate density. Kananyan (1966) showed results for the horizontal circular plate anchors in loose and medium sand. He also performed a series of tests on the inclined anchors and studied the failure on the surface, concluding that most of the soil particles above the anchor moved predominantly in a vertical direction. In these tests, the ultimate uplift load increased in accordance with the inclination angle of the anchors.

In Sergeev and Savchenko (1972), tests were made on models of AP-1 and AP-2 wooden anchor plates having the dimensions 74 cm × 74 cm and 60.5 cm × 60.5 cm and thickness of 5 cm and 10 cm, respectively, as shown in Figs. 2.4 and 2.5. The pressure was measured by gages and arranged one next to the other, converting the entire contact surface of the plate into a continuous metering device. Thirty-six gages were placed on the AP-1 plate, and 25 on the AP-2 plate.

Readings of resistance gages (1) attached to deformable plastic pressure gages were recorded automatically by an AI-1M strain meter. Gages with two indices of compliance of the rigid plate were used in the experiments. In the more compliant gages, with a cross-section of bending plates of approximately 20 mm × 3.2 mm the displacements amounted to

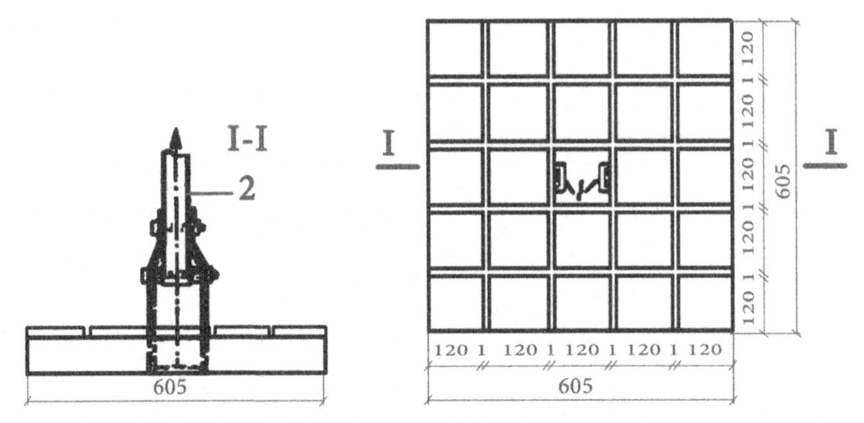

Figure 2.4 AP-1 anchor plate by Sergeev and Savchenko (1972).

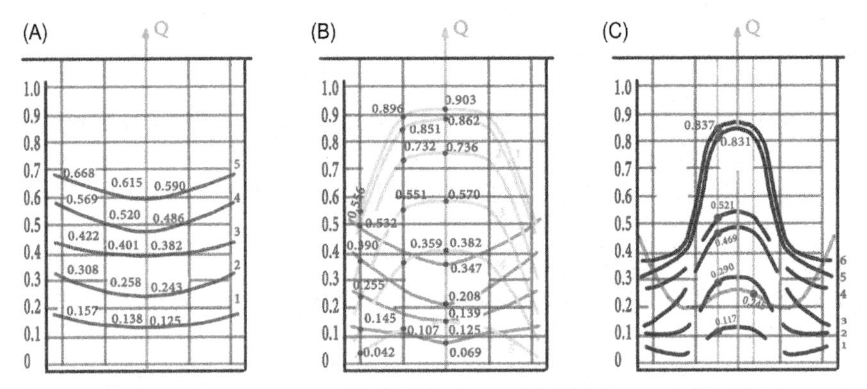

Figure 2.5 Responses pressures: (A) AP-2 in loam; (B) AP-2 in sand; (C) AP-1 in sand; by Sergeev and Savchenko (1972).

3.2–3.3 # per kilogram of loading, and the less compliant gages with a cross-section of the plates of 20 mm × 4 mm, these displacements were 1.8–1.9 #. In most of the experiments, the less compliant gages were placed around the perimeter of the plate model, which corresponded to the character of the theoretical saddle-shaped stress–strain curves obtained in solving the two-dimensional problem. The experiments were conducted in sandy and loamy soil. Prior to the experiments, in order to eliminate the effects of moisture, the model of the plate was wrapped in two layers of polyethylene sheets and heavy cloth, and it was then set in a predesigned foundation pit with a cross-sectional area of 3.5 m × 3.5 m and a depth of 2 m. Medium-grained sand was then placed in the pit or loam. Density was controlled by choice of sample cylinder.

The unit weight of the sand in unconsolidated state was 1.48 g/cm^3, and 1.65 g/cm^3 in the compacted state. The moisture content of the unconsolidated sand was 4.8%, and that of the compacted sand was 4.2%. The standard angle of internal friction of compacted sand in the pressure interval 1–3 kg/cm^2 amounted to 34 degrees; the temerity parameter was 0.02 kg/cm^2, and the strain modulus was $E_0 = 210\ kg/cm^2$.

The bulk weight of loam with layer-by-layer compaction was 1.90 g/cm^3; the moisture content was in the interval 20–23%. The plastic limit and liquid limit lay in the intervals 21–22% and 35–37%, respectively. The standard angle of internal friction was 17 degrees, the cohesion 0.28 kg/cm^2, and the strain modulus $E_0 = 120\ kg/cm^2$. In all experiments with an AP-2 plate, both in unconsolidated sand and in loam, the curves of reaction pressure in the soil had saddle-shaped forms at all stages of loading, including limit load. Repeated loading of the plate introduced no substantial changes in the nature of the distribution of reaction pressures of friable sand or loam. With increased depth of installation of the AP-2 plate in friable sand or loam, the saddle-shaped pressure curves of the soil smoothed out somewhat, which is in qualitative agreement.

Das and Seeley (1975) performed uplift tests on horizontal rectangular anchors ($L/B \leq 5$) in dry sand with a friction angle of $\varnothing = 31$ degrees at a density of 14.8 kN/m^3. For each aspect ratio (L/B), it was found that the anchor capacity increases with the embedment ratio before reaching a constant value at the critical embedment depth. A similar investigation was conducted by Rowe (1978) in dry sand with friction angles $\varnothing = 31–33$ degrees and dry unit weight of $\gamma = 14.9\ kN/m^3$. Polished steel plates were used for the anchors and the interface roughness was measured as $\delta = 16.7$ degrees. Most tests were performed on the anchors with an aspect ratio L/B of 8.75. Rowe (1978) inferred that decreasing the aspect ratio (L/D) leads to the increase of the anchor force (relative to L/B 8.75) of 10, 25, 35, and 120% for L/B ratios of 1–5, respectively.

Thus, the effect of the shape is significant for $L/B \leq 2$ and is of little importance for $L/B > 5$. This suggests that the anchors with aspect ratios of $L/B > 5$ effectively behave as a continuous rectangle and can be compared with the methods which obtain the plane strain conditions. In contrast to the observations of Das and Seeley (1975), Rowe (1978) did not observe a critical embedment depth and the anchor capacity was found to continually increase with the embedment ratio over the range of $H/B = 1–8$.

Extensive chamber testing programs have been studied by Murray and Geddes (1987), who performed the pullout load tests on horizontal strip, circular, and rectangular horizontal anchor plates in dense and medium-dense sand with $\varnothing = 43.6$ degrees and $\varnothing = 36$ degrees, respectively. Anchors were typically 50.8 mm in width or diameter and were tested at aspect ratios (L/B) of 1, 2, 5, and 10.

Based on the observations, Murray and Geddes (1987) reported several conclusions:

1. The pullout load force of rectangular horizontal anchor plates in very dense sand increases with the embedment ratio and with decreasing aspect ratio L/B.
2. There is a difference in the roles of the force of horizontal anchors with rough surfaces compared to those with smooth surfaces (as much as 15%).
3. Experimental results suggest that the behavior of a horizontal anchor plate with an aspected ratio of $L/B = 10$ is similar to that of the strip type and does not differ much from an anchor with $L/B = 5$.
4. The force of the circular horizontal anchor plates in very dense sand is approximately 1.26-times the capacity of square anchors.

Approving these conclusions gives confirmation to all of Rowe (1978)'s findings. It is also of interest to note that for all the tests performed by Murray and Geddes (1987), no critical embedment depth was seen. More recently, Pearce (2000) performed a series of laboratory uplift tests on horizontal circular plate anchors, which were pulled vertically in dense sand. These tests were conducted in a large chamber box.

Various parameters such as the horizontal anchor plate's diameter, uplift rate, and elasticity of the loading system were part of the investigation. The horizontal anchor plates model used for the pullout tests varied in diameter from 50 to 125 mm and were constructed from 8-mm mild steel. Large-diameter anchors were selected (compared with previous researches) due to the identified influence of scale effects on the breakout factor for the anchors of diameters less than 50 mm (Andreadis et al., 1981).

Dickin (1988) performed 41 tests on 25-mm horizontal anchor plates with aspect ratios of $L/B = 1-8$ at embedment ratios H/B up to 8 in both loose and dense sand. A number of conventional gravity tests were also performed and compared to the centrifuge data. The comparison revealed a significant difference in predicting the horizontal anchor plate forces, particularly for square anchors where the force given to the anchor plates by the conventional gravity test was twice of that given by the centrifuge system. Without explaining the reason, Dickin (1988) concluded that the direct extrapolation of the conventional chamber box test resulted at a field scale, would provide overoptimistic predictions of the ultimate force for rectangular horizontal anchor plates the sand. Tagaya et al. (1988) also performed centrifuge testing on rectangular and circular horizontal anchor plates.

Dickin (1988) studied the influence of the anchor geometry, embedment depth and the soil density on the pullout capacity of a 1-m prototype horizontal anchor plate, by subjecting 25-mm models to an acceleration of 40 g in a Liverpool centrifuge. It was found that for the strip anchors, pullout resistance expressed as dimensionless breakout factor, increases significantly with the anchor embedment depth and soil density. However this resistance reduces with the increase in value of the embedment ratio, which is the ratio of length to width of the strip horizontal anchor plate. Failure displacements also increase with the embedment depth but reduce with the soil density and aspect value ratio.

Frydman and Shaham (1989) performed a series of pullout tests on prototype slabs placed at various inclinations and different depths in dense sand. A simple semi-empirical expression is found to reasonably predict the pullout capacity of the continuous, horizontal slab as a foundation for depth–width ratio in their tests. Factors that account for the shape and the inclination are then made, leading to expressions, for the estimation of the pullout capacity of any slab anchor. The following expressions have been proposed for the pullout capacity of the horizontal, rectangular slab horizontal anchor plate in dense sand.

$$(N_q)_r = \left[1 + \frac{D}{B}\tan\varphi\right]\left[1 + \frac{(B/L - 0.15)}{(1 - 0.15)} \times \left(0.51 + 2.35\log\left(\frac{D}{B}\right)\right)\right] \tag{2.1}$$

For loose sand, $D/B \geq 2$

$$(N_q)_r = \left[1 + \frac{D}{B}\tan\varphi\right]\left[1 + 0.5\frac{(B/L - 0.15)}{(1 - 0.15)}\right] \tag{2.2}$$

where $(N_q)_r$ is the pullout capacity.

Ramesh Babu (1998) investigated the pullout capacity and the load deformation behavior of the horizontal shallow anchor plate. Laboratory experiments have been conducted on anchors of different shapes (square, circular, and strip) embedded in medium-dense and dense sands. In addition, the effect of submergence of the soil above horizontal anchor plates has been investigated.

Ghaly (1997a,b) recommended a general expression for the pullout capacity of the vertical anchor plates based statically on the analysis of the experimental test results from the published literature. On similar lines and incorporating appropriate corrections, Ramesh Babu (1998) proposed a general expression for the horizontal anchor plates in the sand by analyzing the results of published experimental data and his own pullout tests data.

For horizontal strip horizontal anchor plate

$$\frac{P_u}{\gamma AD \tan \varphi} = 3.24\left(\frac{D^2}{A}\right)^{0.34} \tag{2.3}$$

For square and circular horizontal anchor plate

$$\frac{P_u}{\gamma AD \tan \varphi} = 3.74\left(\frac{D^2}{A}\right)^{0.34} \tag{2.4}$$

where $\frac{P_u}{\gamma AD \tan \varphi}$, pullout capacity factor and D^2/A is a geometry factor.

Murray and Geddes (1996) presented results of model-scale vertical pulling tests carried out on groups of square anchor plates in row configurations, as illustrated in Fig. 2.6. The tests have been conducted on shallow anchors and embedded in sand. It was shown that the load-displacement was reduced to a common curve, and the load-carrying capacity of a group of anchor plates increased with the spacing between the individual plates up to a limiting critical value. This is how the results of pulling tests with different numbers of anchor plates in a group can be described in a unifying manner. A possible means for predicting the effect of interaction on the uplift capacity of both was suggested. During

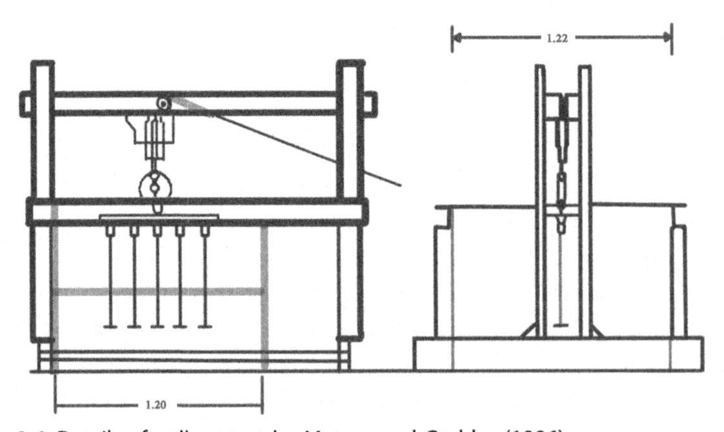

Figure 2.6 Details of pullout test by Murray and Geddes (1996).

laboratory tests on a linear group of five-model and full-scale anchor plates in row configurations, it was shown that the end anchor plates attain the highest load but all loads converge to an equal value as the spacing increases to the critical value, as illustrated in Figs. 2.7 and 2.8.

Only a few investigations, concerning the performance of the ultimate pullout load in cohesion, were recorded in the model studies in the laboratory. One example was provided by Fargic and Marovic (2003) that discussed about the pullout capacity of the anchors in the soil under the applied up-lift force, as illustrated in Fig. 2.9. In the field tests, the pullout forces were gradually increased and the earth surface displacements were measured in two profiles, which were perpendicular to each other. The laboratory and field tests were performed for several embedment depths in the sand of the horizontal anchor plate, diameter ratios in the same sand, and under the same conditions, as illustrated in Fig. 2.10.

Murray and Geddes (1989) investigated the vertical pullout of the horizontal anchor plates in medium-dense sand. The investigation involved the factors in relation to the load-displacement by obtaining the following: size and shape of plate, depth of embedment, sand density and plate surface roughness in the laboratory, as illustrated in Fig. 2.11. The significant differences in behavior were noted between horizontal anchor plates embedded in very dense sand and those embedded in medium-dense sand. Murray and Geddes (1989) also described a research made of the passive resistance of anchorages in sand. The results were compared with previous studies.

The main conclusions of the experimental work were as follows:

1. For the uplift of rectangular plates in very dense sand, the dimensionless load coefficient $P/\gamma AH$ and the corresponding displacement at failure increase with an increase of H/B ratio and a decrease of L/B ratio. The dependence of uplift resistance on L/B is described in terms of a shape factor. There is also a marked increase in $P/\gamma AH$ and the corresponding displacement in very dense sand for plates with high surface friction angle compared to plates with polished surfaces.
2. Significant differences in behavior were noted between plates embedded in very dense sand and those embedded in medium-dense sand. While the dimensionless load coefficient $P/\gamma AH$ is greater in very dense sand, the corresponding displacements are considerably less. Of particular concern is the recorded behavior of circular plates in medium-dense sand where large abrupt decreases in uplift resistance were recorded prior to the absolute maximum uplift resistance.

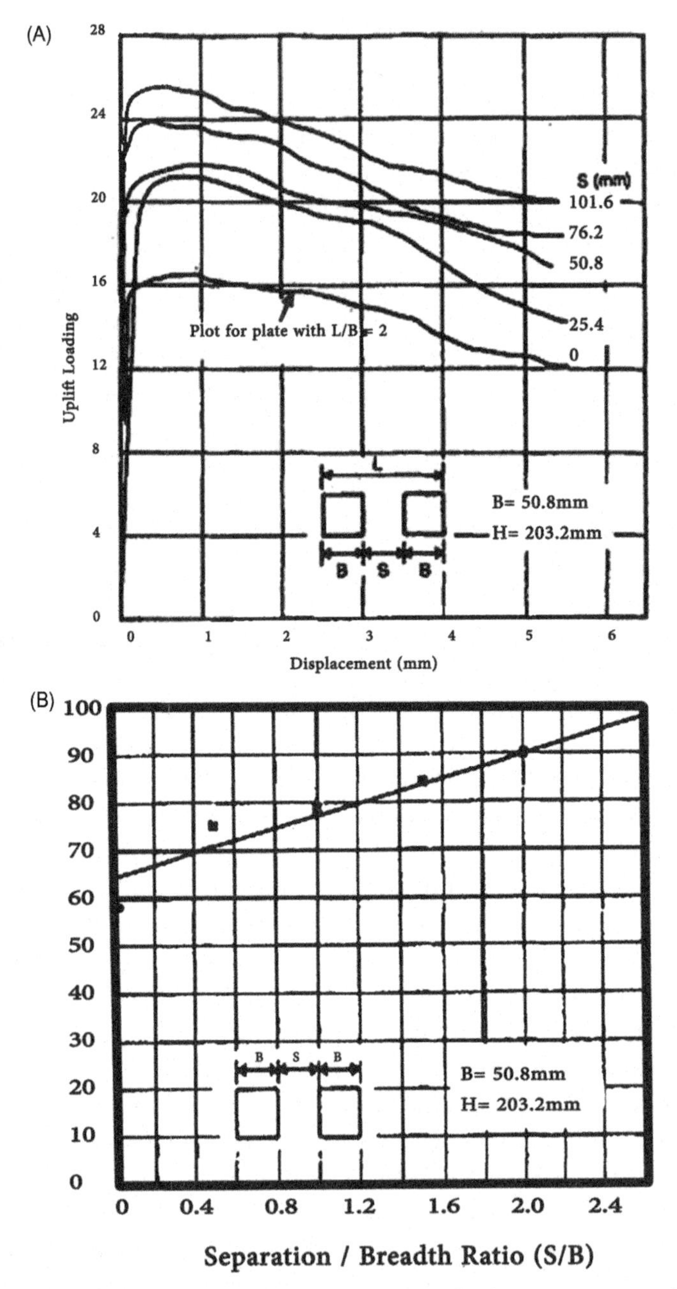

Figure 2.7 Square anchor plates results. (A) Load–displacement curves; (B) group efficiency results by Murray and Geddes (1996).

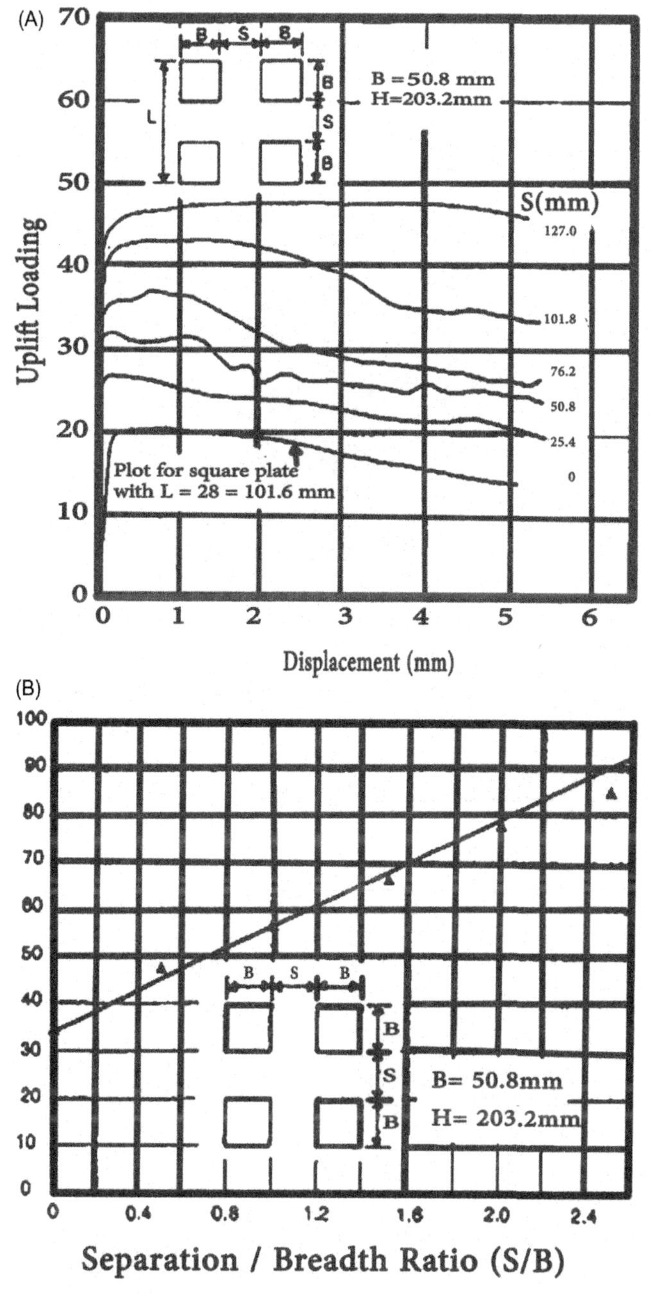

Figure 2.8 Four square anchor plates. (A) Load–displacement curves; (B) group efficiency results by Murray and Geddes (1996).

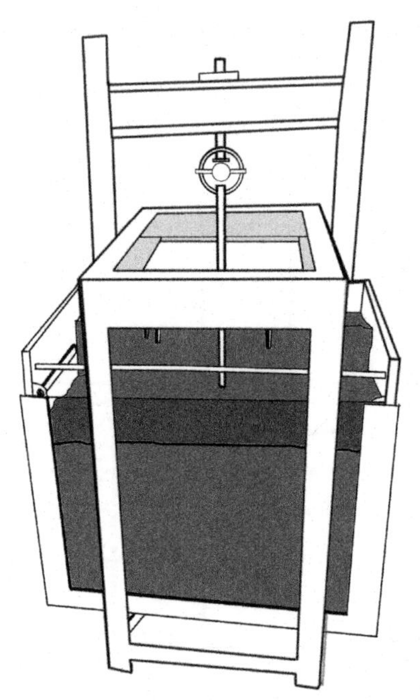

Figure 2.9 Laboratory test devices by Fargic and Marovic (2003).

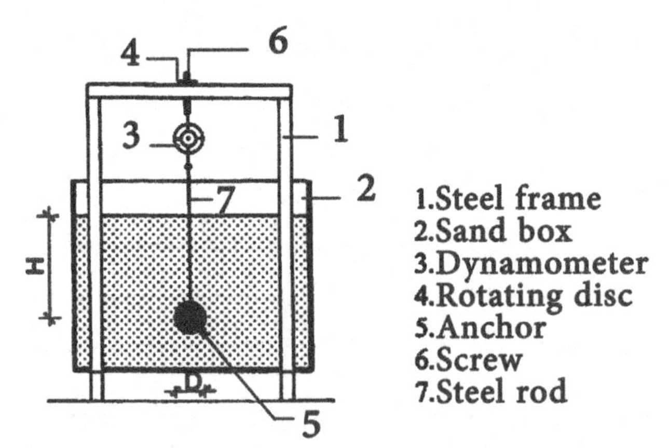

Figure 2.10 Scheme of the laboratory test for the pullout test by Fargic and Marovic (2003).

3. For circular plates in very dense sand, there appears to be a consistent relationship for both the dimensionless load coefficients P/γAH and the corresponding displacements, for all plates tested, when plotted against H/D. Similar relationships do not appear to exist in medium

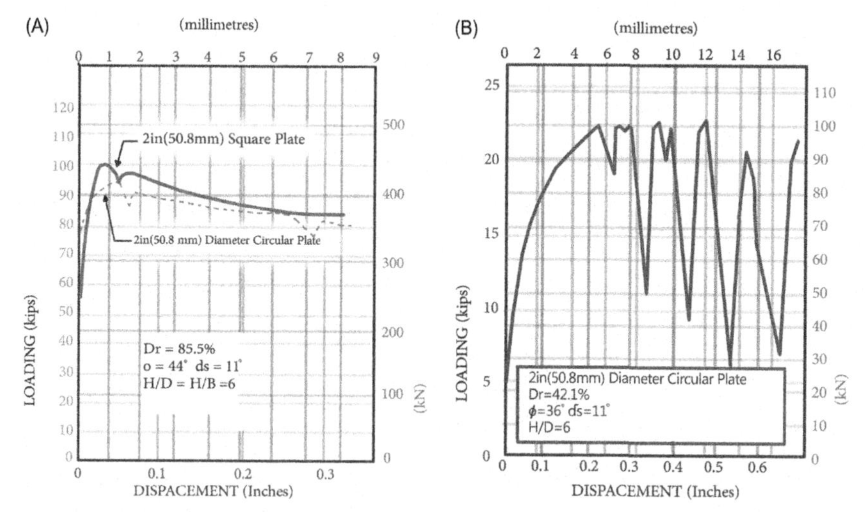

Figure 2.11 Load/displacement curves of Murray and Geddes (1989). (A) Very dense sand; (B) medium-dense sand.

dense sand. In very dense sand, the $P/\gamma AH$ values for circular plates are, on average, approximately 1.26-times those of square plates for $H/B = H/D$.

Dickin and Laman (2007) investigated the physical modeling of the horizontal anchor plates in a centrifuge. The centrifuge incorporated balanced swinging buckets, which were 0.57 m long, 0.46 m wide, and 0.23 m deep. Physical research investigated the pullout response of 1-m-wide strip anchors in sand, the results of which indicated that maximum resistances increase with the anchor embedment ratio and sand packing. The physical research showed that the breakout factors for 1-m wide strip anchors increased with the anchor embedment ratio and the sand packing.

Kumar and Bhoi (2008) used a group of multiple strip anchors placed in the sand and subjected them to equal magnitudes of vertical upward uplift loads that had been determined by model experiments as illustrated in Fig. 2.12. Instead of using a number of horizontal anchor plates in the experiments, a single horizontal anchor plate was used by modeling the boundary conditions along the plane of symmetry on both sides of the horizontal anchor plate. The effect of interference, due to a number of multiple strip horizontal anchor plates placed in a granular medium at different embedment depths, was investigated by conducting a series of small-scale-model tests. The experiment data

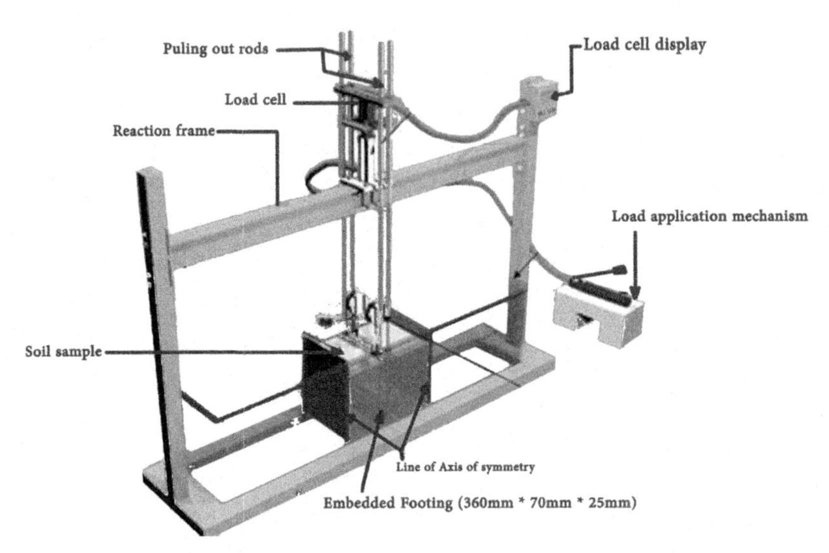

Figure 2.12 3D view of the chosen experimental setup by Kumar and Bhoi (2008).

clearly reveal that the magnitude of the failure load reduces quite extensively with a decrease in the spacing between the horizontal anchor plates (Tables 2.1 and 2.2).

Dickin (1988) investigated the influence of anchor geometry, embedment ratio, and soil density on the uplift capacity of 1-m prototype anchors. Uplift resistances expressed as dimensionless breakout factors increase significantly with anchor embedment ratio and soil density but reduce with increased embedment ratio, as illustrated in Fig. 2.13. Failure displacements also increase with embedment ratio but reduce with increased soil density and aspect ratio. The influence of anchor geometry is relatively insensitive to anchor size but increases with both embedment ratio and soil density. In general, the design approaches considered underestimate the observed capacity of 1-m strip anchors in dense sand, although predictions using the Meyerhof and Adams (1968), Rowe and Davis (1982) theories are acceptable. However, both these approaches appear overoptimistic for anchors in loose-packed soil. In this case, Vesic's theory gives the closest agreement, while the formula of Majer yields overly conservative designs. Pullout capacities for isolated anchors may be obtained from the strip values in combination with the empirical shape factors reported in this research.

Kumar and Bhoi (2008) the ultimate uplift resistance of a group of multiple strip anchors placed in sand and subjected to equal magnitudes

Table 2.1 Previous experimental works on horizontal square anchor plates in cohesionless soil

Anchor	Type of testing	Anchor shape	Anchor size	Friction angle	*L/D*
Sergeev and Savchenko (1972)	Chamber	Square	74 mm 60.5 mm	34°	0.74–2.2
Das and Seeley (1975)	Chamber	Square	51 mm	31°	1–5
Rowe (1978)	Chamber	Square	51 mm	32°	1–8
Ovesen (1981)	Centrifuge and field	Square	20 mm	29.5° and 37.7°	1–3.39
Dickin (1988)	Centrifuge and chamber	Square	25 mm 50 mm	38–41° Loose 48–51° Dense	1–8
Murray and Geddes (1989)	Chamber	Square	50.8 mm	36° Loose 43.6° Dense	1–8
Sarac (1989)	–	Square	–	37.5° and 48°	0.35–4
Murray and Geddes (1989)	Chamber	Square	50.8 mm	43.6° Dense	1–8
Ramesh Babu (1998)	Chamber	Square	50, 75, 100 mm	32.5° Loose 45° Dense	1–8

Table 2.2 Previous experimental works on horizontal strip anchor plates in cohesionless soil

Anchor	Type of testing	Anchor shape	Anchor size	Friction angle	*L/D*
Das and Seeley (1975)	Chamber	Strip	51 mm	31°	1–5
Rowe (1978)	Chamber	Strip	51 mm	32°	1–8
Murray and Geddes (1987)	Chamber	Strip	50.8 mm	44° and 36°	1–10
Frydman and Shamam (1989)	Chamber and field	Strip	19 mm 200 mm	30° Loose 45° Dense	2.5–9.35
Dickin (1988)	Centrifuge and Chamber	Strip	25 mm 50 mm	38–41° Loose 48–51° Dense	1–8
Tagaya et al. (1988)	Centrifuge	Strip	15 mm	42°	3–7.02
Murray and Geddes (1989)	Chamber	Strip	50.8 mm	36° Loose 43.6° Dense	1–8
Dickin (1994)	Centrifuge	Strip	25 mm	38–41° Loose 48–51° Dense	1–7
Ramesh Babu (1998)	Chamber	Strip	50, 75, 100 mm	32.5° Loose 45° Dense	1–8
Dickin and Laman (2007)	Centrifuge	Strip	100–250 mm	35° Loose 51° Dense	1–8
Kumar and Bhoi (2008)	Chamber	Multiple strip	70–370 mm	37.4°, 41.8°, 44.8°	0.41–12.86

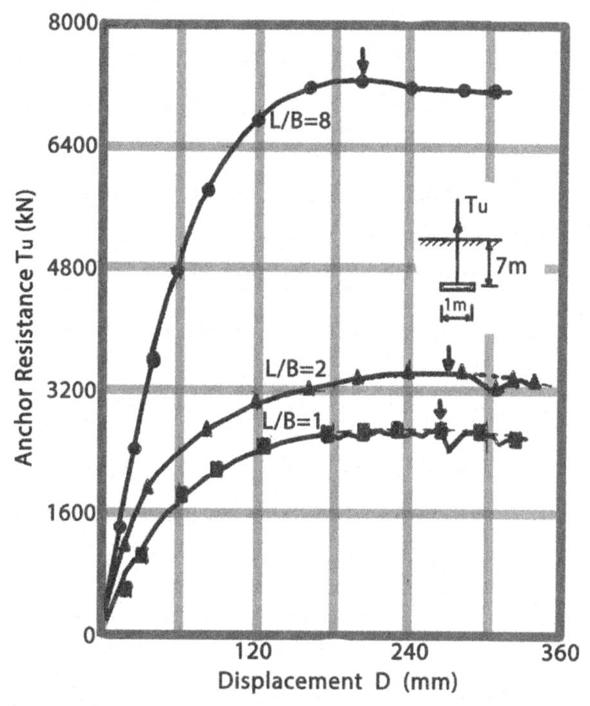

Figure 2.13 Effect of anchor plates in dense sand by Dickin (1988).

of vertical upward pullout loads has been determined by means of model experiments, as illustrated in Fig. 2.14. Instead of using a number of anchor plates in the experiments, a single anchor plate was used by simulating the boundary conditions along the planes of symmetry on both the sides of the anchor plate. The effect of clear spacing(s) between the anchors, for different combinations of embedment ratio (λ) of anchors and friction angle ($\varnothing$) of soil mass, was examined in detail. The results were presented in terms of a one-dimensional efficiency factor which was defined as the ratio of the failure load for an intervening strip anchor of a given width (B) to that of a single strip anchor plate having the same width. It was clearly noted that the magnitude of efficiency factor reduces quite extensively with a decrease in the spacing between the anchors. The magnitude of efficiency factor for a given s/B was found to vary only marginally with respect to changes in $\varnothing$ and λ, as illustrated in Fig. 2.15. The experimental results presented in this research compare reasonably well with the theoretical and experimental data available in the literature.

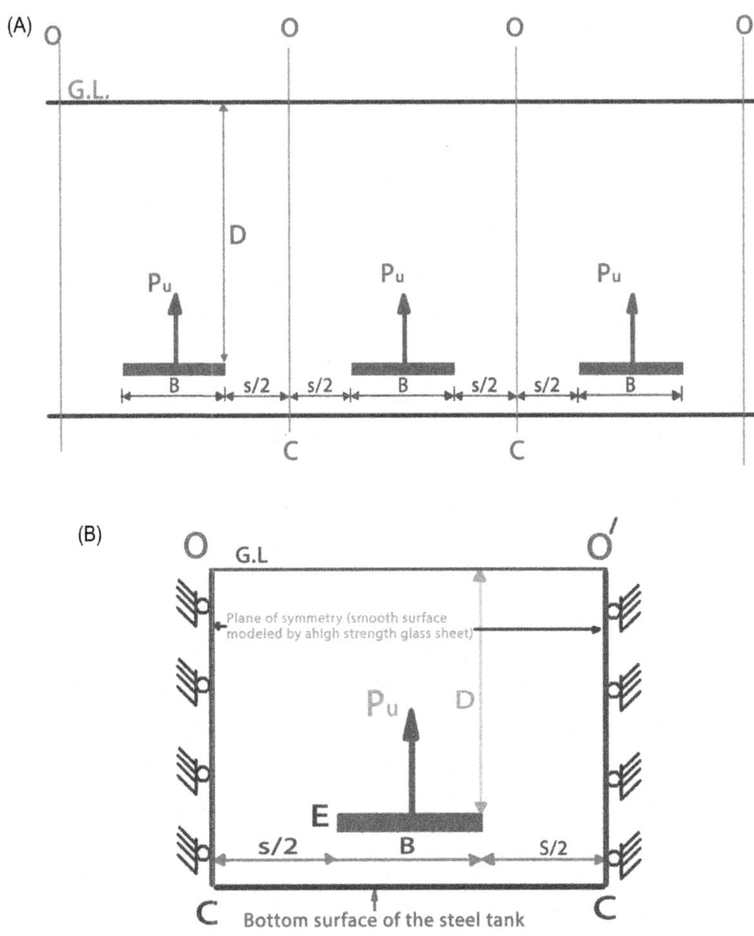

Figure 2.14 (A) Problem definition; and (B) domain for carrying out the experiments by Kumar and Bhoi (2008).

In the literature, empirical relationships were suggested to correlate the uplift capacity of a single circular anchor to its geometrical properties and the characteristics of the surrounding soil (Hoyt & Clemence, 1989; Mitschand & Clemence, 1985; Udwari, Rodgers, & Singh, 1979). Furthermore, analytical models were developed to predict the pullout capacity of single anchors in sand based on an assumed failure mechanism, made of a set of planes (Ghaly, Hanna, & Hanna, 1991; Ilamparuthi & Muthukrishnaiah, 1999; Murray & Geddes, 1987), circular arc surface (Baker & Kondner, 1966; Balla, 1961; Vesic, 1971), and cylindrical surface (El Hansy, 1980; Meyerhof & Adams, 1968; Mitsch & Clemence, 1985).

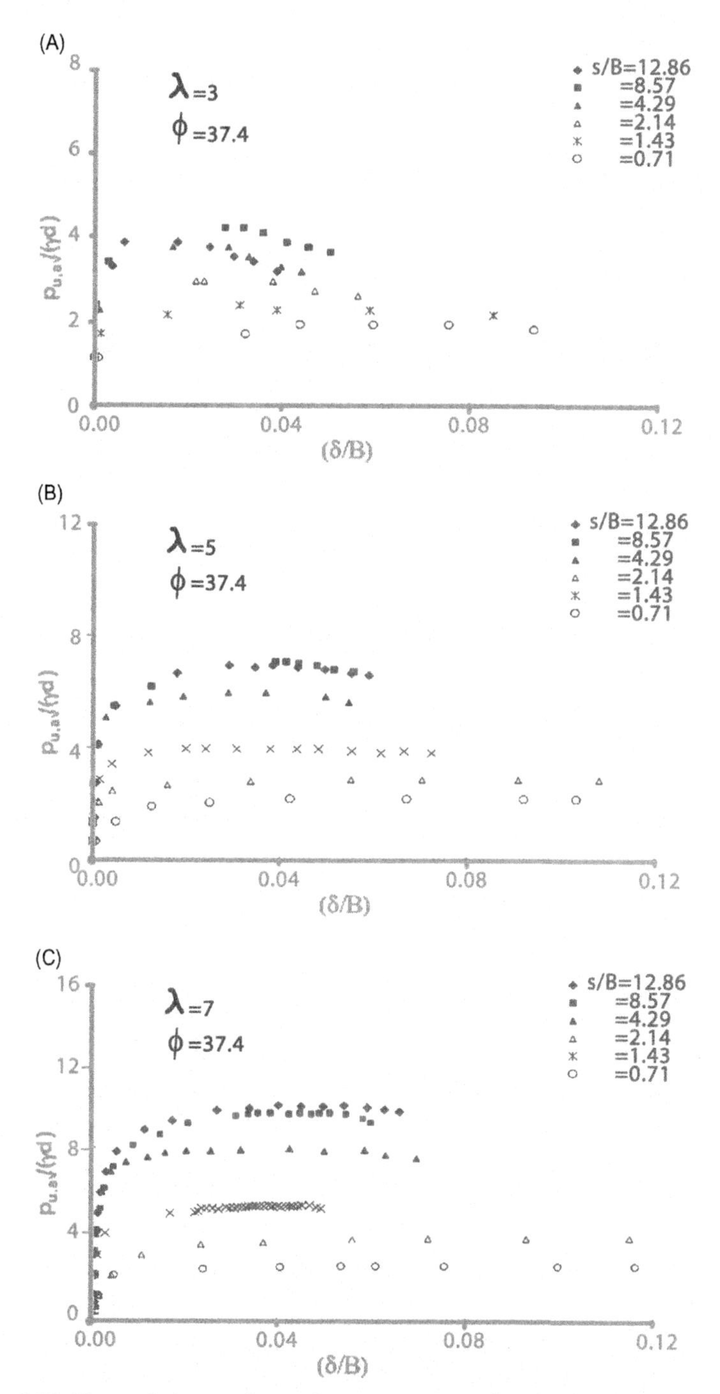

Figure 2.15 The variation pullout ultimate capacity for different values of (*s*/*B*) for ∅ = 37.4 at (A) λ = 3; (B) λ = 5; and (C) λ = 7 by Kumar and Bhoi (2008).

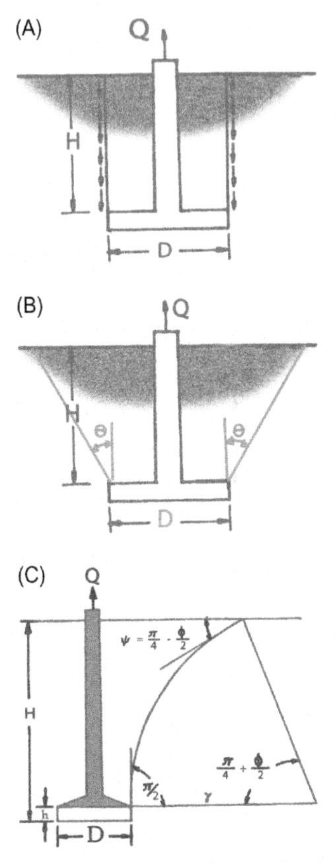

Figure 2.16 Assume failure surfaces: (A) cylinder method, (B) cone method, (C) Balla method.

Wang and Wu (1980) developed an upper bound solution based on a failure mechanism consisting of a straight line and a log-spiral, as illustrated in Fig. 2.16. Ghaly and Hanna (1994) developed theoretical models to predict the pullout capacity of single vertical anchors installed into shallow, transition, and deep depths using the limit equilibrium technique together with Kotter's differential equation to calculate the shear stresses acting on the log-spiral surface. Dickin and Leung (1983) and Ilamparuthi, Dickin, and Muthukrisnaiah (2002) have presented a thorough review of the design theories for circular anchors available in the literature.

Ilamparuthi and Muthukrishnaiah (1999) investigated uplift capacities of anchor plates to support offshore structures in great water depths, as

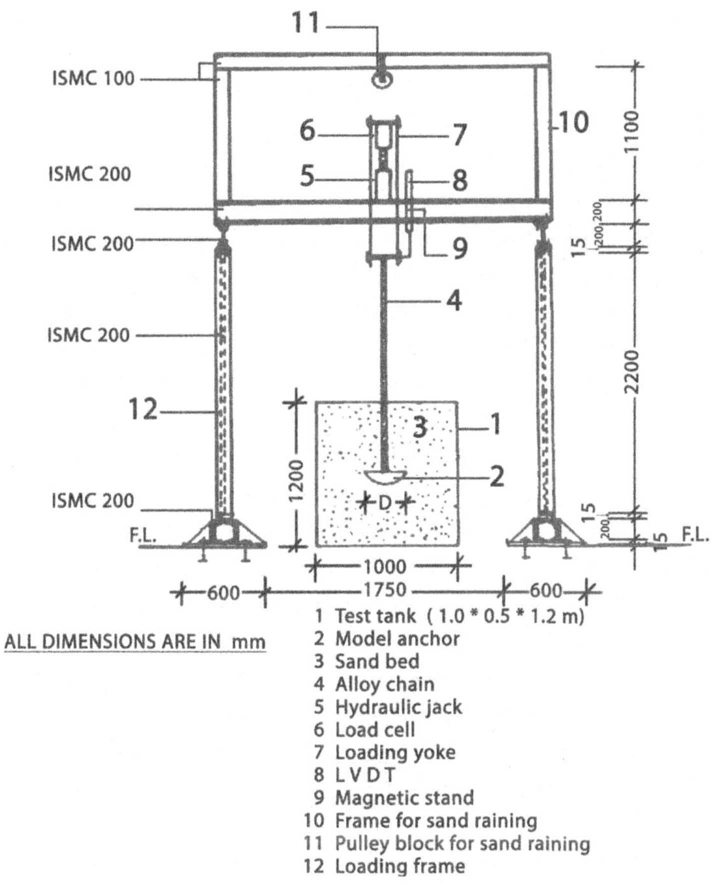

Figure 2.17 Schematic diagram of pullout test by Ilamparuthi and Muthukrishnaiah (1999).

illustrated in Fig. 2.17. The capacities of plate and mushroom type anchors are generally estimated based on the shape of rupture surface, as illustrated in Fig. 2.18. An attempt has been made in the present investigation to delineate the rupture surfaces of anchors embedded in submerged and dry sand bed sat various depths. The results exhibited two different modes of failure depending on the embedment ratio, namely, shallow and deep anchor behavior. The load–displacement curves exhibited three- and two-phase behaviors for shallow and deep anchors, respectively. Negative pore water pressures recorded in submerged sand also exhibited variation similar to that of pullout load versus anchor

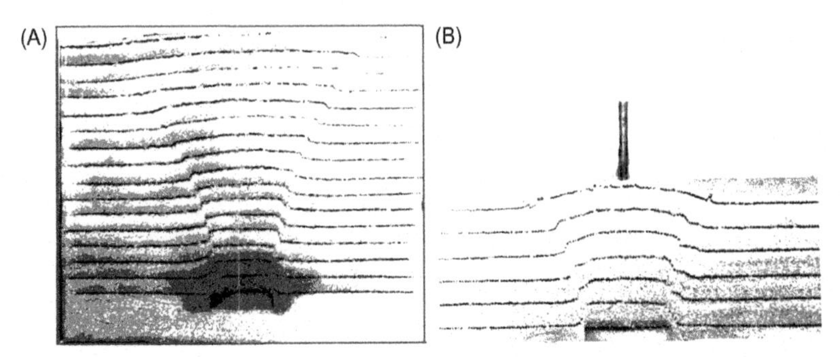

Figure 2.18 Failure pattern of anchor plate with various embedded depths. (A) $L = 201$ mm, (B) $L = 527$ mm in dense sand by Ilamparuthi and Muthukrishnaiah (1999).

displacement. Table 2.3 showed the previous experimental works on horizontal circular anchor plates in cohesionless soil.

From the experimental investigations carried out on half-cut models of flat circular and curved anchors embedded at different depths in dry and submerged sand beds of different densities, the following conclusions are arrived at:

1. Two types of rupture surface were observed, depending on the embedment ratio, one emerging to the surface of the sand bed and the other confined with in the sand bed, irrespective of shape of anchor, density of sand bed and dry or submerged conditions. A transition exists between the two types of failure, giving rise to the concept of the "critical embedment ratio."
2. For embedment ratios less than the critical embedment ratio (shallow anchors), the rupture surface is a gentle curve, convex upwards, which can be closely approximated to a plane surface. The plane surface makes an angle of $\varnothing/2 \pm 2$ degrees with the vertical irrespective of density, submergence of sand bed and shape of anchor.
3. For embedment ratios greater than critical embedment ratio (deep anchors), the rupture surface emerging from the edge of the anchor makes an angle of 0.8 degrees with the vertical irrespective of the density of the sand bed and the rupture surface is confined with in the sand bed.
4. The load versus displacement relationship is different for shallow and deep anchors. Three-phase behavior for shallow anchors and

Table 2.3 Previous experimental works on horizontal circular anchor plates in cohesionless soil

Anchor	Type of testing	Anchor size	Friction angle	*L/D*
Balla (1961)	Chamber	60–120 mm	38°	1–6
Kananyan (1966)	Field	400 mm	37°	8.56
Hanna and Carr (1971)	Chamber	38 mm	37°	4–11.2
Hanna, Sparks, and Yilmaz (1971)	Chamber and Field	38 and 150 mm	37°	4–11.2
Kwasniewski, Sulikowska, and Walker (1975)	Field	700 mm	28°	2.71
Fadl (1981)	Chamber	75 mm	36.5° Loose 41.5° Dense	4–8
Andreadis et al. (1981)	Chamber	80–150 mm	37° and 42.5°	1–14
Ovesen (1981)	Centrifuge and Field	20 mm	29.5° and 37.7°	1–3.39
Murray and Geddes (1987)	Chamber	50.8 mm	44° and 36°	1–10
Tagaya et al. (1988)	Centrifuge	15 mm	42°	3–7.02
Sarac (1989)	Chamber	–	37.5°, 48°	0.35–4
Bouazza and Finlay (1990)	Chamber	37.5 mm	33.8°, 39°, 43.7°	2–5
Sakai and Tanaka (1998)	Chamber	30–200 mm	–	1–3
Ramesh Babu (1998)	Chamber	50, 75, 100 mm	32.5° Loose 45° Dense	1–8
Ilamparuthi and Muthukrisnaiah (1999)	Chamber	100–400 mm	33.5° Loose 38.5° Medium 43° Dense	1–8
Pearce (2000)	Chamber	50–125 mm	Loose to very dense	1–15
Liu, Li, Yang, Zhang, and Liu (2010)	Field	50.8 mm	30.8° Loose 43.3° Dense	1–3

two-phase behavior for deep anchors were observed irrespective of density and submergence of the sand bed.

5. For the densities studied in the submerged sand bed, the measured pore water pressures were negative. The variation of negative pore water pressure with anchor displacement is similar to that of pullout load versus anchor displacement.

Liu, Liu, and Gao (2010) reported a test on a cohesionless soil displacement field during anchor uplifting. The cohesionless soil displacement is calculated using the digital image correlation (DIC) method from two images: one is taken at the initial stage and the other is at the peak pullout load moment, as illustrated in Fig. 2.19. The failure pattern is identified by locating the maximum shear strains deduced from the dense conditions: (1) in loose sand, the shearing bands start at the edges of the anchor plate an coverage and from a bell above the anchor; (2) while in dense sand these shear bands extend outward to the ground surface with an inclination angle with the vertical of approximately ½ to ⅓ of the friction angle of the soil.

Symmetrical anchor plates are a foundation system that can resist tensile load with the support of the surrounding soil in which the symmetrical anchor plate is embedded. It is used by soil structures as a structural member, primarily to resist uplift loads and overturning moments and to ensure the structural stability. A wide variety of soil anchor systems (plate, irregular shape anchor, grouted, and helical anchors) have been developed in order to satisfy the increase in foundations to resist the uplift responses. Engineers and authors proved that the uplift response can be improved by grouping the symmetrical anchor plates, increasing the unit weight, the embedment ratio, and the size of symmetrical anchor plates.

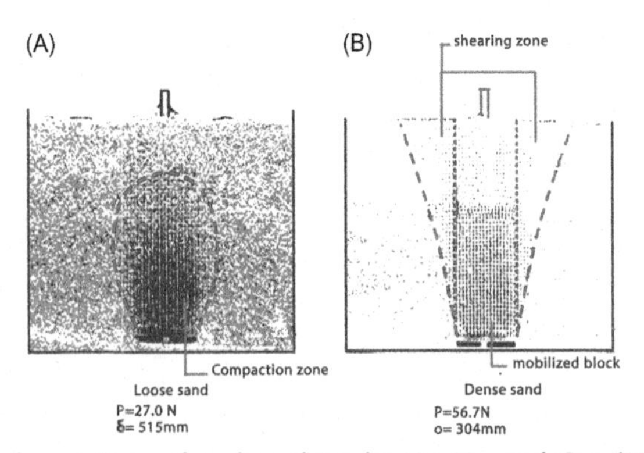

Figure 2.19 Failure pattern of anchor plates by Liu, Liu, and Gao (2010). (A) Loose sand and (B) dense sand.

Innovations in geosynthetics as reinforcement materials in the field of civil engineering have provided possible solutions in symmetrical anchor plate responses. Research into the uplift response of symmetrical anchor plates embedded in non-reinforced soil has been reported by Bala (1961); Meyerhof and Adams (1968); Vesic (1971, 1972); Hanna et al. (1971); Meyerhof (1951); Neely, Stuart, and Graham (1973); Baset (1977); Das and Seeley (1975); Rowe and Davis (1982); Saran, Ranjan, and Nene (1986); Dickin (1987) although research in the area of symmetrical anchor plates embedded in reinforced soil such as Johnston (1984); Subbarao, Mukhopadhyay, and Sinha (1988); Singh (1992); Rajagopal and SriHari (1996) were less extensive.

Selvadurai (1989) investigated the performance of geogrids for anchoring 150-mm diameter and 850-mm-long pipelines embedded in sand. Andreadis et al. (1981), Ghaly et al. (1991), Krishnaswamy and Parashar (1992), and Ilamaparuthi and Muthukrishnaiah (2001) were the few known researchers that worked in reinforced sand under uplift tests. Krishnaswamy and Parashar (1992) investigated the experimental test in a reinforced sand bed and others in non-reinforced sand beds. More extensive experimental research was carried out (Selvadurai, 1993) to evaluate the treatment of a 215-mm-diameter pipe with a length of 1610 mm embedded in reinforced sand beds. The inclusion of geogrids immediately above the pipeline in an inclined setup increased the uplift response by about 80%. The uplift response of symmetrical anchor plates embedded in sand bed with geosynthetic reinforcement material was studied by Krishnaswamy and Parashar (1992).

Krishnaswamy and Parashar (1994) investigated the uplift response of symmetrical anchor plates such as circular anchor plates (60 mm in diameter) and rectangular plates (53 mm wide with lengths varying from 23.8 to 53 mm) embedded in clay and sand with and without geosynthetics. Placing the geosynthetics directly on the symmetrical anchor plate was proved to be beneficial in achieving maximum increase in the uplift response, although they found that two layers of geogrid reinforcement does not increase the uplift capacity predominantly. Ilamparuthi and Dickin (2001) investigated the treatment of belled pile anchors in a reinforced sand bed and formulated a hyperbolic method for calculating the breakout factor. Ravichandran and Ilamparuthi (2004) evaluated the treatment of rectangular anchor plates in non-reinforced and reinforced cohesionless soil beds.

Kingshri, Ilamparuthi, and Ravichandran (2005) evaluated two series of experimental tests to understand the influence of stiffness and opening size of geosynthetic reinforcement materials on the uplift capacity of rectangular anchor plates. The first series of tests were on a combination of geocomposite and geogrid, and the second series of tests were on two layers of geogrid as reinforcements and concluded that the performance of geocomposite and geogrid (two layer) combination was found to be more effective in resisting uplift response than two combined layers of geogrids.

The performance of symmetrical anchor plates has been studied by many researchers in non-reinforced sand, but relatively little is known about the performance of symmetrical anchor plates in a reinforced soil bed. Johnston (1984) investigated the pullout response of geogrids.

Subbarao et al. (1988) evaluated the improvement in pullout load using geotextiles as ties to symmetrical anchor plates embedded in sand. Experimental tests were conducted on reinforced concrete model symmetrical anchor plates of cylindrical and belled shape, with polypropylene ties of width 55 mm and thickness 0.72 mm being used. Results showed that symmetrical anchor plates with geotextile ties offered greater uplift responses than those without ties. Furthermore, the use of single layers of ties close to the symmetrical anchor plates were reported to be more effective than the use of multiple layers. Failure mechanisms for various uplift design methods can be found in Fig. 2.20.

Increasing the use of anchors to resist and sustain uplift forces may be achieved by increasing the size and depth of an anchor or the

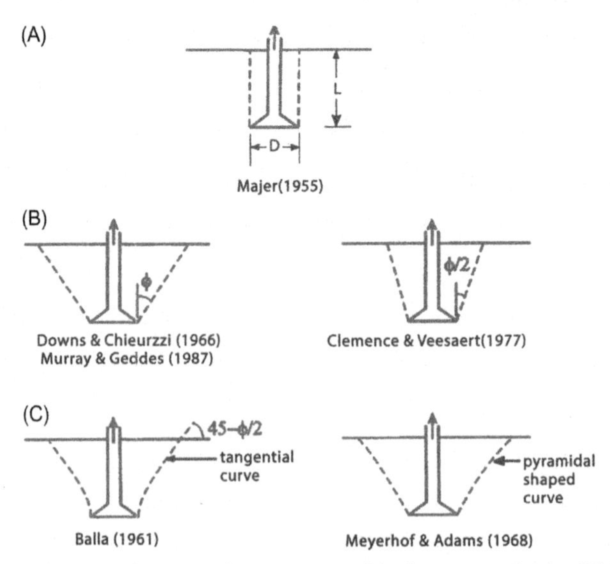

Figure 2.20 Failure mechanisms for various uplift design methods. (A) Vertical slip surface model, (B) inverted truncated cone model, and (C) curved slip surface model.

improvement of soil in which these anchors are embedded, or both. In restricted situations, increasing the size and depth of an anchor may not be economical compared with other alternatives. However, soil improvement can be attained by the inclusion of soil reinforcement to resist larger uplift forces. However, few investigations on the behavior of horizontal plates in reinforced soil beds under uplift loads have been reported.

Subbarao et al. (1988) studied the improvement in pullout capacity by using geotextiles as ties to reinforced concrete model anchors embedded in sand. Selvadurai (1989, 1993) reported significant enhancement, of the order of 80–100%, in the uplift capacity of pipelines embedded in fine and coarse-grained soil beds reinforced by inclusion of geogrids immediately above the pipeline in an inclined configuration. Krishnaswamy and Parashar (1992, 1994) studied the uplift behavior of circular plates and rectangular plates embedded in cohesive and cohesionless soils with and without geosynthetic reinforcement and reported that the geocomposite reinforcement offered higher uplift resistance than both geogrid and geotextile reinforcement.

Ilamparuthi and Dickin (2001) investigated the behavior of soil reinforcement on the uplift response of piles embedded in sand through a model test in which a cylindrical gravel-filled geogrid cell was placed near the pile base, as illustrated in Fig. 2.21. The authors reported increases in the uplift response of piles with many factors such as diameter of the geogrid cell, sand density, pile bell diameter, and embedment, as illustrated in Fig. 2.22.

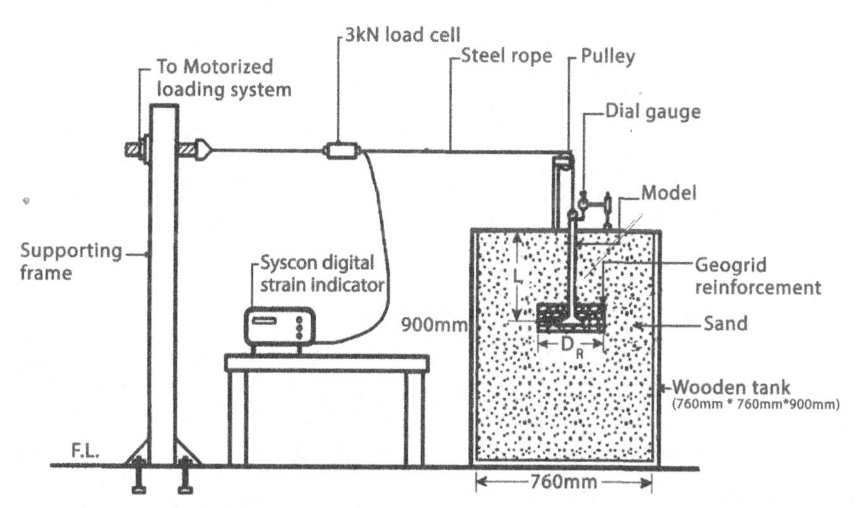

Figure 2.21 Schematic diagram of experimental test by Ilamparuthi and Dickin (2001).

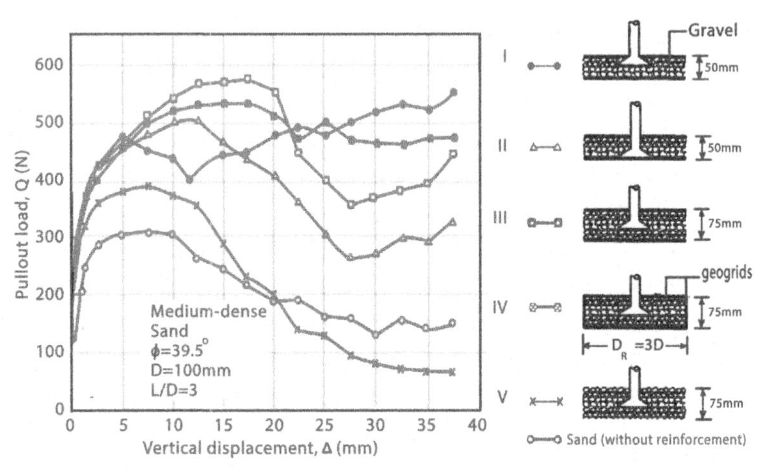

Figure 2.22 Non-reinforced and reinforced results of symmetrical anchor plates under uplift test by Ilamparuthi and Dickin (2001).

El Sawwaf (2007) conducted a laboratory experimental investigation on a strip anchor plate to investigate the uplift response in sand. Uplift response of symmetrical anchor plates located close to sandy slopes with and without geosynthetic reinforcement has been evaluated in tests. Strip anchor plates were used in the experimental work to research the effect of soil reinforcement on the uplift behavior of anchor plates using plane strain. Strip anchor plates made of steel of 498 mm in length, 6.0 mm in thickness, and 80 mm in width were made with a special hole 3.0 mm in diameter in the center and used in the research. The authors found many conclusions, such as the fact that increasing the ultimate pullout response of an anchor plate embedded to the slope crest and anchor plate improvement is very dependent on geosynthetic layer length and increases significantly until the amount beyond which further increase in the layer length does not show a significant contribution to the anchor resistance. Geosynthetic layers were placed to reinforced slope, as shown in Fig. 2.23.

Ilamparuthi, Ravichandran, and Mohammed Toufeeq (2008) investigated two series on the submerged sand effect of symmetrical anchor plates on uplift response of treated non-reinforced and reinforced sand, as illustrated in Fig. 2.24. Influence factors such as embedment ratio, density and number of geofrid layers were investigated. Authors were reported the load—displacement behavior of non-reinforced and reinforced sand for a given density and embedment ratio were similar except higher peak and residual loads due to the geogrid materials. In the case of cyclic load, the displacement of symmetrical anchor plate was increased with

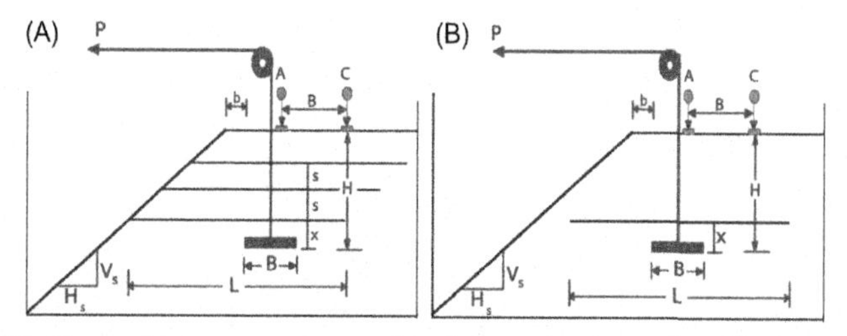

Figure 2.23 Geometric parameters of reinforced slope by El Sawwaf (2007). (A) Multi-Layers (B) Single Layer.

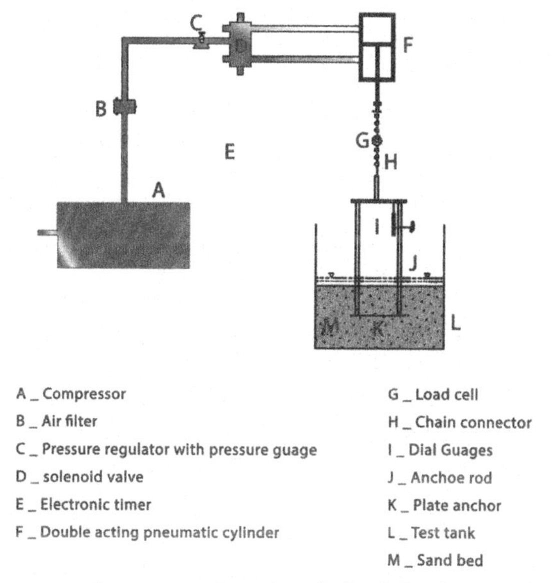

Figure 2.24 Test set up for monotonic and cyclic loads by Ilamparuthi et al. (2008).

decreasing rate and reached almost a constant value after 350 cycles, as illustrated in Fig. 2.25.

2.2.1 Limitations of Previous Experimental Works

As a conclusion, the history shows that different researches have been performed regarding horizontal anchor plate performance in the sand. These researches used different horizontal anchor plates and soil parameters. Inevitably such a wide range of parameters will contribute to conflicting conclusions for the ultimate pullout load of the horizontal anchor plates.

Some of the works did not include in their presentation the internal friction angle, anchor roughness, and anchor size. However, most authors

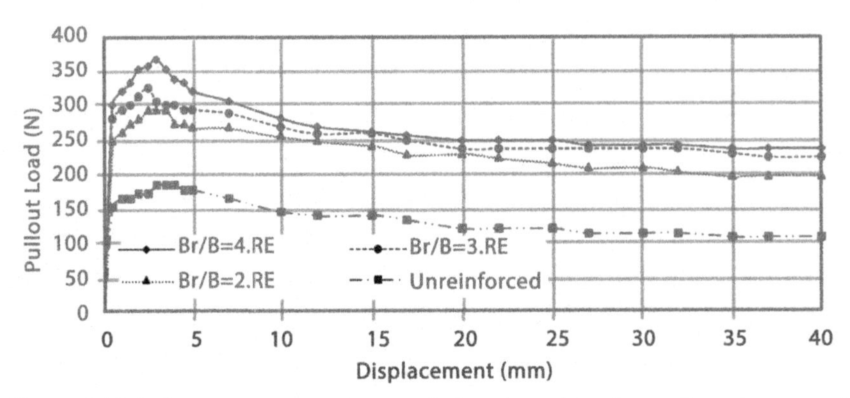

Figure 2.25 Pullout response on symmetrical anchor plates in medium dense sand bed for $H/B = 4$ by Ilamparuthi et al. (2008).

obtained their internal friction angle using the direct shear test or triaxial compression test. Unfortunately, the results obtained from the laboratory tests are typically a specific problem and are difficult to extend and develop to field problems due to the different materials or the geometric parameters in the field scale. The most soil anchor works have been concerned with the uplift problem on embedded in non-reinforced soils under pullout test. Symmetrical anchor plates are a foundation system that can resist tensile load with the support of the surrounding soil in which the symmetrical anchor plate is embedded. Engineers and authors proved that the uplift response can be improved by grouping the symmetrical anchor plates, increasing the unit weight, embedment ratio, and the size of symmetrical anchor plates. Innovations in geosynthetics in the field of geotechnical engineering as reinforcement materials have been found to be possible solutions in symmetrical anchor plate responses. Unfortunately the importance of reinforcement in submergence has received very little attention by researchers.

2.3 REVIEW OF PREVIOUS NUMERICAL WORKS

This section presents theoretical and numerical studies of horizontal anchor plates from last 50 years until now. Numerical methods for estimating the ultimate pullout capacity of anchor plates have been developed. This part discusses different theories and numerical studies in anchor plates by previous researchers. Analyses, beginning from Meyerhof and Adams (1968) until the most recent analyses such as

Kuzar and Kumar (2009) are reviewed. Numerical analysis and laboratory works were pioneered by Vesic (1971), Sarac (1989), Smit (1998), Krishna (2000), Fargic and Marovic (2003), Merifield and Sloan (2006), Dickin and Laman (2007), Kumar and Bhoi (2008), Kuzer and Kumar (2009) are also discussed. The results include the latest theories and numerical studies which were more numerous than experimental works. The ultimate pullout capacities of horizontal anchor plates in sand were investigated based on previous theories and numerical studies conducted. Numerical analysis and laboratory works were performed to research the performance of ultimate pullout loading anchor plates. However, there are relatively few documents in the technical literature which deal with an anchor plate in sand. For future analysis purposes, the limiting ultimate capacity of anchor plate embedded into sand can be assessed based on numerical and experimental work.

During the last 50 years, various researchers have conducted laboratory studies to better understand and predict the ultimate uplift loading of anchors in a range of soil types. In contrast to the variety of experimental results already discussed, very few rigorous numerical analyses have been studied to determine the pullout loading of anchor plates in soil. In the literature, several theories have been proposed to calculate the pullout force of anchors, the difference between each one of them being mainly in the shape of the selected failure surface. It is essential to verify theoretical solutions/numerical analysis with experimental studies wherever possible; laboratory test results are problem specific. This is important to do in the case of geomechanics which deals with a highly nonlinear material which often displays pronounced scale effects. Thus, it is often difficult to extend the findings from small-scale laboratory research to full-scale projects with different materials or geometric parameters. Existing numerical analyses generally selected a condition of plane strain for the case of a continuous strip anchor plate or axi-symmetry for the case of circular anchors. Numerical methods have been developed to estimate the ultimate pullout capacity of anchor plates that have been developed. One of the earliest publications concerning ultimate pullout capacity of anchor plates was by Mors (1959), which proposed a failure surface in the soil at ultimate load which may be approximated as a truncated cone having an apex angle α equal to (90 degrees + Ø/2), as shown in Fig. 2.26. The net ultimate pullout capacity was assumed to be equal to the weight of the soil mass bounded by the sides of the cone and the shearing resistance over the failure area surface was ignored.

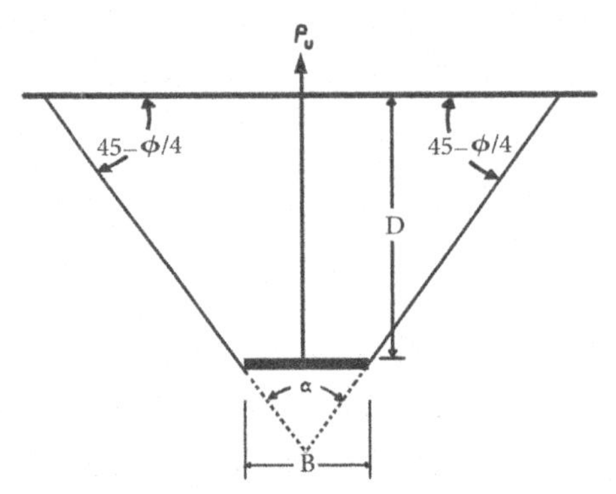

Figure 2.26 Failure surface assumed by Mors (1959).

$$P_u = \gamma V \tag{2.5}$$

where, V, volume of the soil in the truncated and γ, unit weight of soil.

Downs and Chieurzzi (1966), based on similar theories, investigated the hypothesis that the apex angle is always equal to 60 degrees, irrespective of friction angle of the soil. However, Teng (1962) and Sutherland (1988) found that this assumption might be unsafe in same cases common with increase in depth.

An approximate semi-empirical theory for the pullout loading force of horizontal strip, circular, and rectangular anchors has been proposed by Meyerhof and Adams (1968). For a strip anchor, an expression for the ultimate capacity was selected by considering the equilibrium of the block of soil directly above the anchor (i.e., contained within the zone made when vertical planes are extended from the anchor edges). The capacity was assumed to act along the vertical planes extending from the anchor shape, while the total passive earth pressure was assumed to act at some angle to these vertical planes. This angle was selected based on laboratory test results while the passive earth pressures were evaluated from the results of Caquot and Kerisel (1948). For shallow anchor plates where the failure surface develops to the soil surface, the ultimate capacity was determined by considering equilibrium of the material between the anchor and soil surface. The magnitude of H was determined from the observed extent of the failure surface from laboratory works.

The analysis of strip footings was developed by Meyerhof and Adams (1968) to include circular anchor plates using a semi-empirical shape factor to modify the passive earth pressure obtained for the plane strain case. The failure surface was assumed to be a vertical cylindrical surface through the anchor edge and extending to the soil surface. An approximate analysis for the capacity of rectangular anchor plates was selected as for downward loads (Meyerhof, 1951), by assuming that the ground pressure along the circular perimeter of the two end portions of the failure surface was governed by the same shape factor assumed for circular anchors.

It was, however, based on two key adoptions: namely, the edge of the failure surface and the distribution of stress along the failure surface. Even so, the theory presented by Meyerhof and Adams (1968) has been found to give reasonable estimates for a wide range of anchor plate problems. It is one of only two methods available for appraising the force of rectangular anchor plates.

Clemence and Veesaert (1977) showed a formulation for shallow circular anchors in sand assuming a linear failure making an angle of $\beta = \varnothing/2$ with the vertical through the shape of the anchor plate, as shown in Fig. 2.27. The contribution of shearing resistance along the length of the failure surface was approximately taken into consideration by selecting a suitable value of ground pressure coefficient from laboratory model works. The net ultimate capacity can be given as

$$P_u = \gamma V + \pi \gamma K_o \tan \varnothing \cos^2\left(\frac{\varnothing}{2}\right)\left(\frac{BD^2}{2} + \frac{D^3 \tan \frac{\varnothing}{2}}{3}\right) \tag{2.6}$$

where, V is the volume of the truncated cone above the anchor and K_o is the coefficient of lateral earth pressure; they suggested that the magnitude of K_o may vary between 0.6 and 1.5 with an average value of about 1.

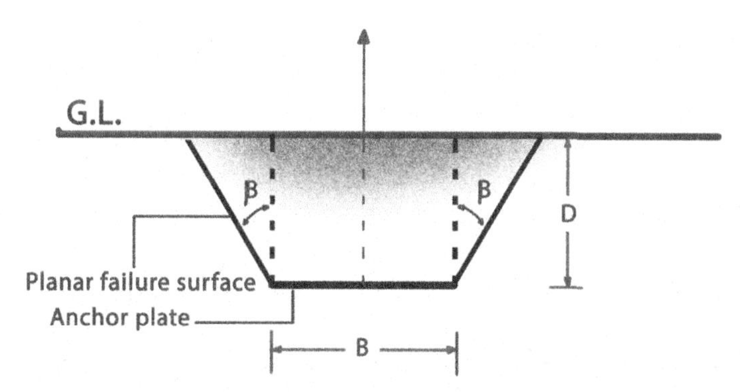

Figure 2.27 Failure surface assumed by Clemence and Veesaert (1977).

The finite element method had also been used by Vemeer and Sutjiadi (1985), Tagaya, Tanaka, and Aboshi (1983), Tagaya et al. (1988), and Sakai and Tanaka (1998). Unfortunately, only limited results were presented in these researches.

Tagaya et al. (1983, 1988) conducted two-dimensional plane strain and axi-symmetric finite element analyses using the constitutive law of Lade and Duncan (1975). Scale effects for circular anchor plates were investigated by Sakai and Tanaka (1998) using a constitutive model for a non-associated strain hardening-softening elastoplastic material in dense sand. The effect of shear band thickness was also introduced.

Koutsabeloulis and Griffiths (1989) investigated the trapdoor problem using the initial stress finite element method. Both plane strain and axi-symmetric researches were conducted. The researchers concluded that an associated flow rule has little effect on the collapse load for strip anchor plates but has a significant effect (30%) on circular anchors. Large displacements were observed for circular anchor plates prior to collapse.

In the limit equilibrium method (LEM), an arbitrary failure surface is adopted along with a distribution of stress along the selected surface. Equilibrium conditions are considered for the failing soil mass and an estimate of the collapse load is assumed. In the research of horizontal anchor force, the failure mechanism is generally assumed to be log spiral in edge (Ghaly & Hanna, 1994; Murray & Geddes, 1987; Saeedy, 1987; Sarac, 1989) and the distribution of stress is obtained by using either Kotter's equation (Balla, 1961), or by using an assumption regarding the orientation of the resultant force acting on the failure plane.

Upper and lower bound limit analysis techniques have been used studied by Murray and Geddes (1987, 1989), Basudhar and Singh (1994), and Smith (1998) to estimate the capacity of horizontal and vertical strip anchor plates. Basudhar and Singh (1994) selected estimates using a generalized lower bound procedure based on finite elements and non-linear programing similar to that of Sloan (1988). The solutions of Murray and Geddes (1987, 1989) were selected by manually constructing cinematically admissible failure mechanisms (upper bound), while Smith (1998) showed a novel rigorous limiting stress field (lower bound) solution for the trapdoor problem. Only a few numbers of investigations into the performance of ultimate pullout loading in numerical studies in sand were recorded. For example, it is Fargic and Marovic (2003) that discussed the pullout capacity of anchor plates in soil under applied vertical force. Computation of the pullout and uplift force was performed by the finite element method. For a gravity load, the concept of initial stresses in Gauss points was selected. In the first increment of computation, these stresses were added to the vector of total stress. The soil was modeled by an elastoplastic constituent material model and the associated flow rule was used. The soil mechanics parameters were determined by standard tests conducted on disturbed samples. For a complex constitutive numerical model of material to describe an actual state of soil, a greater number of soil mechanics parameters must be available. The tensile strength of the soil materials was crucial only in a few cases, and the problem of tensile anchor plates is one of them. An iterative procedure was used as the first procedure. The elements with tensile stresses were excluded from the following steps by diminishing the different modulus. More sophisticated constitutive laws are required for an exact analysis, and an adequate finite element method code program has to be prepared.

Merifield and Sloan (2006) used much numerical solution for analysis of anchor plates. Until this time very few rigorous numerical analyses had been performed to determine the pullout capacity of anchor plates in sand, as illustrated in Fig. 2.28. Although it is essential to verify theoretical solutions/numerical analysis with experimental studies wherever possible, results selected from their laboratory testing alone were typically problem specific. This was particularly the case in geotechnics, where they were dealing with a highly non-linear material that sometimes displays pronounced scale effects, as illustrated in Figs. 2.29 and 2.30.

As the cost of performing laboratory works on each and every field problem combination is prohibitive, it is necessary to be able to model soil pullout loading numerically for the purposes of design. Generally, existing numerical analyses assumed a condition of plane strain for the

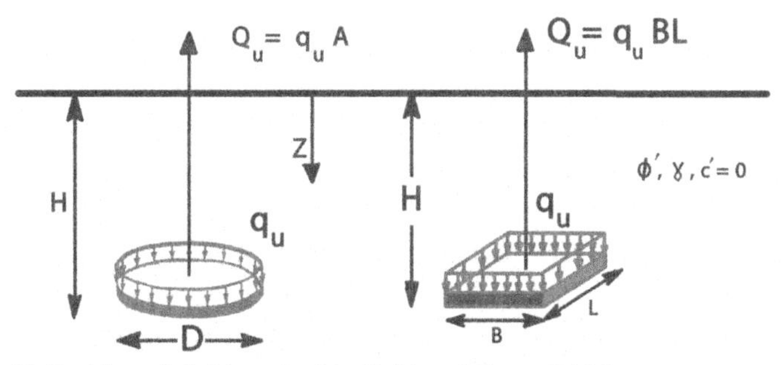

Figure 2.28 Problem definitions by Merifield and Sloan (2006).

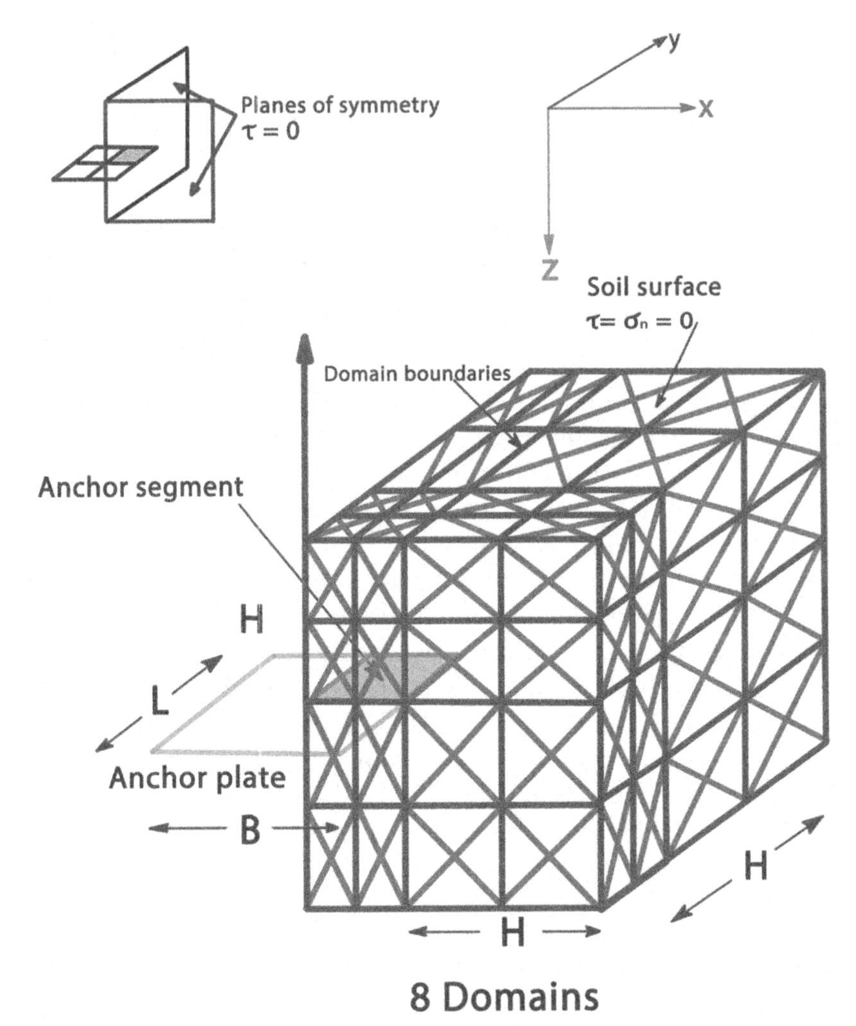

Figure 2.29 Mesh for square anchor plate by Merifield and Sloan (2006).

case of a continuous strip anchor plate or axi-symmetry for the case of circular anchor plates. The researchers were unaware of any three-dimensional numerical analyses to ascertain the effect of anchor plate shape on the uplift capacity.

Dickin and Laman (2007) investigated the numerical modeling of anchor plates by PLAXIS, a finite element program. In the numerical modeling, the anchor plates were 0.57 m long, 0.46 m wide, and 0.23 m deep. Numerical analysis research investigated the uplift response of 1-m-wide strip anchors in sand where the results indicated that maximum

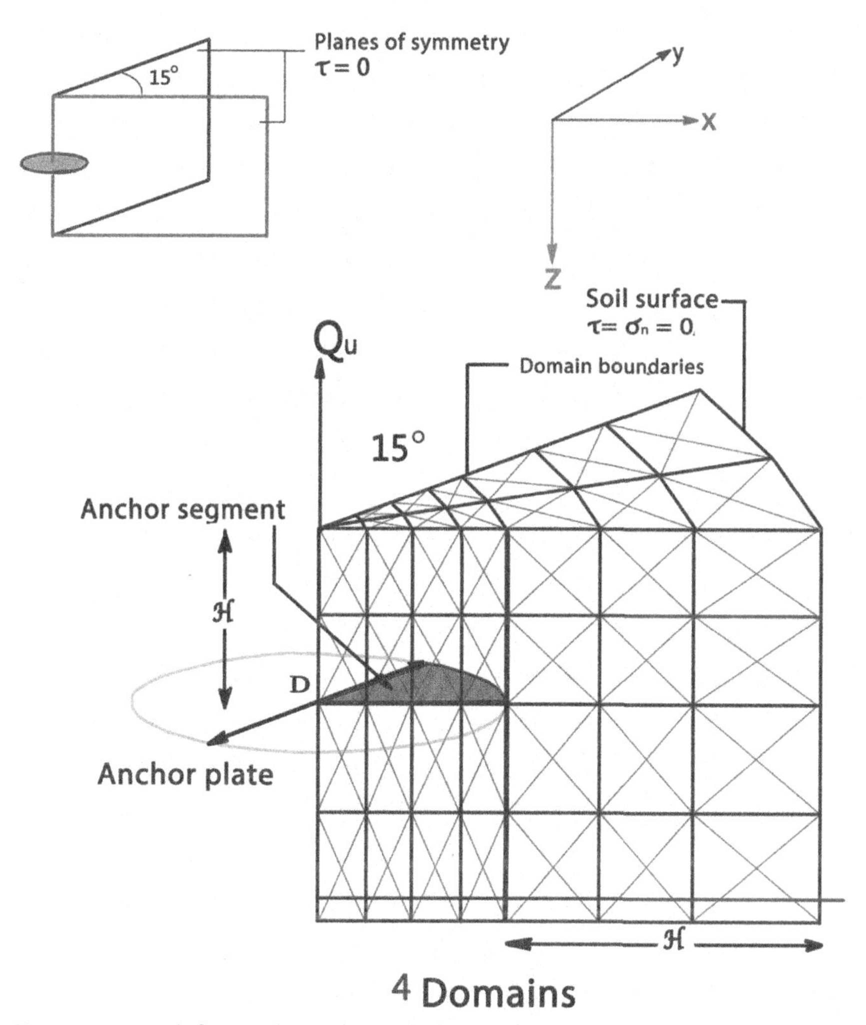

Figure 2.30 Mesh for circular anchor plate by Merifield and Sloan (2006).

ultimate capacity increased with anchor embedment ratio and sand packing. The research was carried out using a plane strain model for anchors in both loose and dense sand. During the generation of the mesh, 15-node triangular elements were obtained in the determination of stresses as illustrated in Figs. 2.31 and 2.32.

Kumar and Bhoi (2008) used a group of multiple strip anchor plates placed in sand and subjected them to equal magnitudes of vertical upward pullout loads determined by numerical solutions, as illustrated in

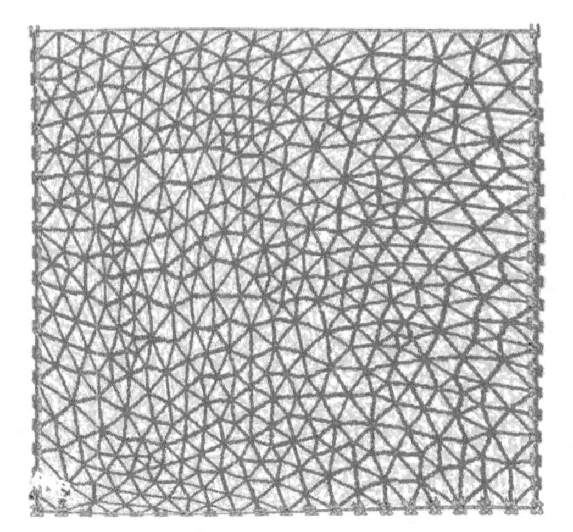

Figure 2.31 Very fine mesh in analysis of a strip anchor with an embedment ratio $L/D=7$ in PLAXIS by Dickin and Laman (2007).

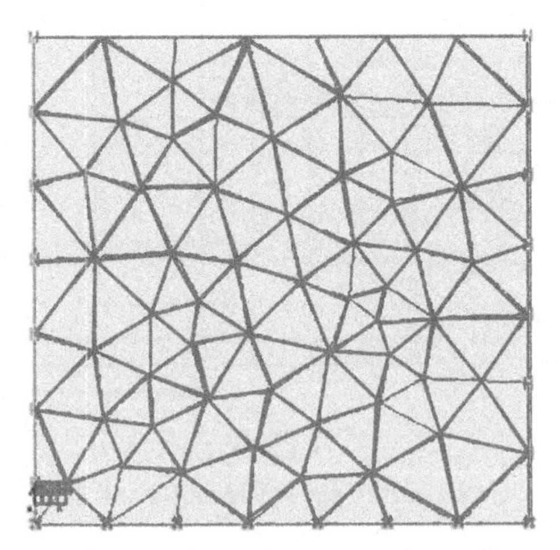

Figure 2.32 Coarse mesh in analysis of a strip anchor with an embedment ratio $L/D=7$ in PLAXIS by Dickin and Laman (2007).

Fig. 2.33. Instead of using a number of anchor plates in numerical modeling, a single anchor plate was used by modeling the boundary conditions along the plane of symmetry on both the sides of the anchor plate. The effect of interference due to a number of multiple strip anchor plates placed in a granular medium at different embedment depths was investigated by conducting a series of small numerical models.

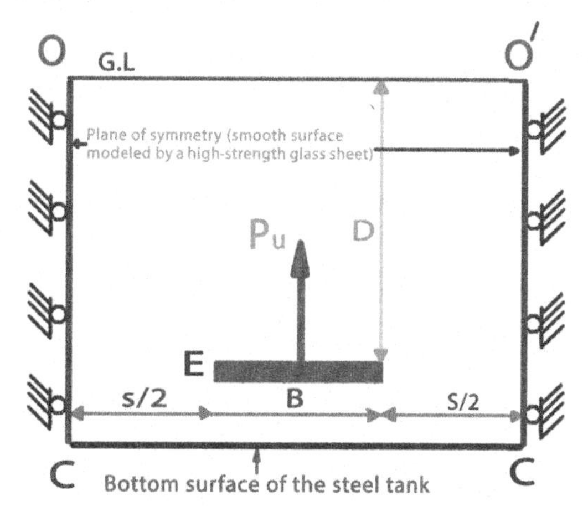

Figure 2.33 Definition of the problem by Kumar and Bhoi (2008).

Kuzer and Kumar (2009) used a group of two spaced strip anchor plates. The vertical pullout loading of two interfering rigid rough strip anchors embedded horizontally in sand as illustrated in Fig. 2.34. The analysis was performed by obtaining an upper bound theorem of limit analysis combination with finite element and linear programing. The authors used an upper bound finite element limit analysis; the efficiency factor ξ_γ was computed for a group of two closely spaced strip anchor plates in sand. Previous numerical analyses on horizontal square anchor plates in cohesionless soils can be found in Table 2.4.

Kouzer and Kumar (2009) determined the vertical uplift capacity of an infinite numbers of rigid strip plate anchors placed in a group and embedded in sand, as illustrated in Fig. 2.35. An upper bound theorem of limit analysis in combination with finite element sand linear programing was used to perform the analysis. All the anchors were loaded to failure simultaneously with the same magnitudes of the failure loads. The ratio of failure load for a spacing S between the anchors, the magnitude of the efficiency factor interfering strip anchor, for given width B and depth d, to that of an isolated strip anchor having the same width and depth. They have been found to reduce substantially with a decrease in the spacing between the anchors. It was noted that a rigid soil magnitude of wedge just above the anchor plate and bounded by planar rupture surfaces separates out and it moves with a velocity the same as that of the anchor plate itself as illustrated in Fig. 2.36. Previous numerical analyses on horizontal strip anchor plates in cohesionless soils can be found in Table 2.5.

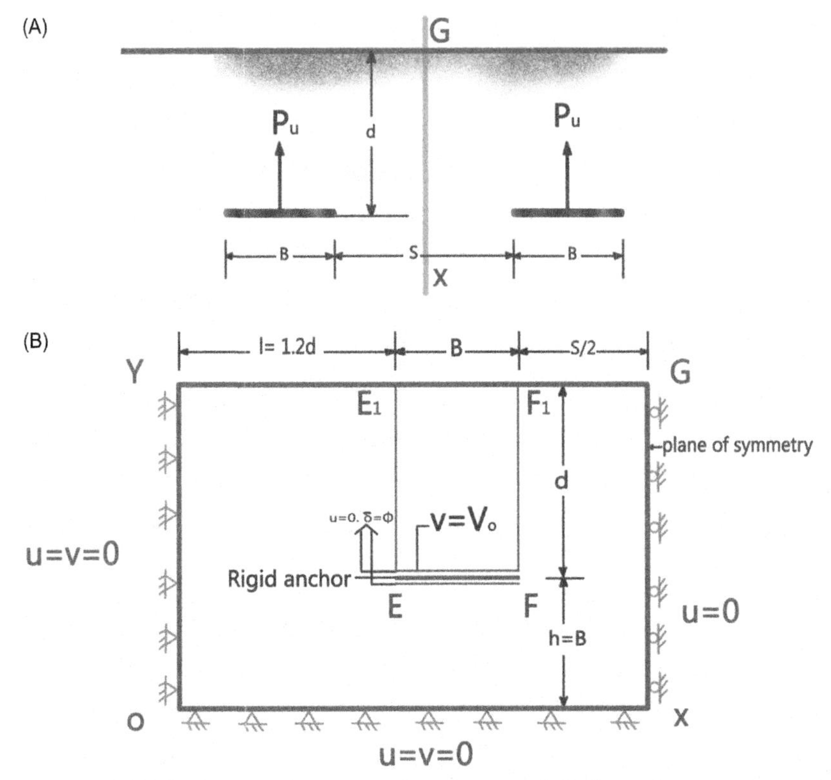

Figure 2.34 (A) Position and loading of anchors, and (B) boundary conditions by Kuzer and Kumar (2009).

Table 2.4 Previous numerical analyses on horizontal square anchor plates in cohesionless soils

Researcher	Analysis method	Friction angle	*L/D*
Meyerhof and Adams (1968)	Limit equilibrium	—	—
Murray and Geddes (1987)	Limit analysis and limit equilibrium	All	All
Sarac (1989)	Limit equilibrium	0–50°	1–4
Merifield and Sloan (2006)	Limit analysis	20–40°	1–10

The results available from the analysis were found to compare reasonably well with the available theoretical and experimental data from the literature. By performing an upper bound limit analysis in combination with finite elements, the vertical uplift resistance is obtained for an

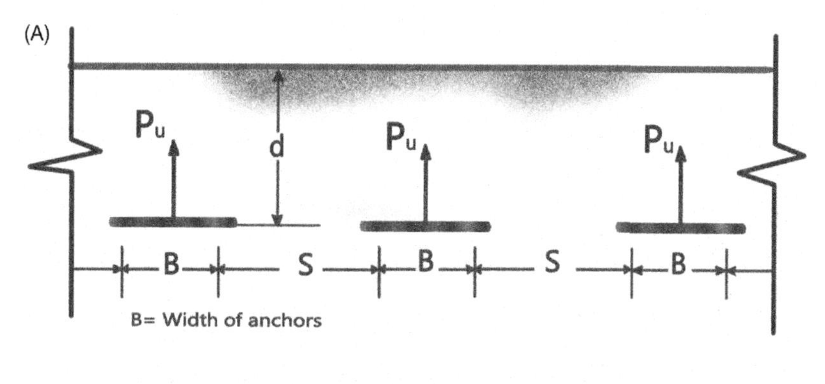

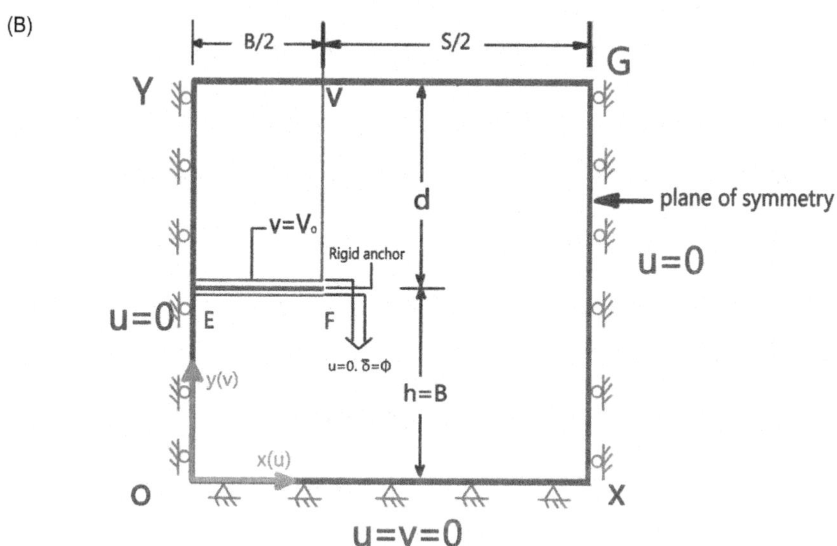

Figure 2.35 (A) Position of anchor plate, (B) boundary condition by Kouzer and Kumar (2009).

infinite number of equally spaced rough strip anchors placed in sand. It was noted that when the clear spacing S between the anchors is approximately greater than $2d$ tan, hardly any interference between the adjoining anchors exists. On the other hand, for $S < d$ tan, the uplift resistance of anchors decreases substantially with a decrease in the spacing between the anchors, and the value of becomes minimum at $S/B = 0$. For given values of S/B and, the magnitude of becomes smaller with an increase in Ø. From the observed nodal velocity patterns, it was seen that in all the cases the soil mass lying above the anchor moves almost as a single

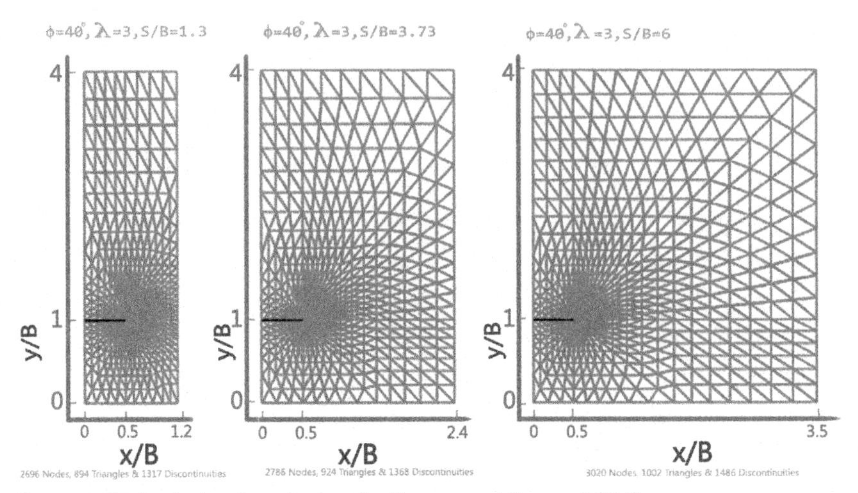

Figure 2.36 Analysis of mesh sizes by Kouzer and Kumar (2009).

rigid unit with a velocity the same as that of the anchor itself. The numerical solution obtained in this part provides a reasonable comparison with the available experimental and theoretical data from the literature.

Numerical methods in estimating the ultimate pullout capacity of plate anchors have been developed. One of the earliest publications concerning ultimate pullout capacity of anchor plates was by Mors (1959) which proposed a failure surface in the soil at ultimate load which may be approximated as a truncated cone having an apex angle α equal to (90 degrees + Ø/2), as shown in Fig. 2.37. The net ultimate pullout capacity was assumed to be equal to the weight of the soil mass bounded by the sides of the cone and the shearing resistance over the failure area surface was ignored.

$$P_u = \gamma V \tag{2.7}$$

where, V, volume of the soil in the truncated and γ, unit weight of soil.

This information makes guidance for the design and evaluation of anchor systems used to prevent the sliding and/or overturning of laterally loaded structures founded in soils. The typical system of forces acting on a simple anchor are shown in Fig. 2.38. The pullout force is given by the typical equation

$$P_u = P_s + W + P_t \tag{2.8}$$

Table 2.5 Previous numerical analyses on horizontal strip anchor plates in cohesionless soils

Researcher	Analysis method	Anchor roughness	*L/D*
Meyerhof and Adams (1968)	Limit equilibrium	–	–
Vesic (1971)	Cavity expansion	–	0–5
Rowe and Davis (1982)	ElastoplasticFinite element	Smooth	1–8
Vemeer and Sutjiadi (1985)	ElastoplasticFinite element	–	1–8
Tagaya et al. (1983, 1988)	ElastoplasticFinite element	–	0–30
Murray and Geddes (1987)	Limit analysis and limit equilibrium	–	All
Koutsabeloulis and Griffiths (1989)	Finite element method—initial stress	–	1–8
Basudhar and Singh (1994)	Limit analysis	Rough Smooth	1–8
Kanakapura, Rao, and Kumar (1994)	Method of characteristics	Smooth	2–10
Smith (1998)	Limit analysis	–	1–28
Krishna (2000)	Finite difference method	–	1–8
Dickin and Laman (2007)	Finite element method	–	1–8
Kumar and Bhoi (2008)	Finite element method	–	0.41–12.86
Kuzer and Kumar (2009)	Limit analysis and displacement finite element method	Rough	0–5.03

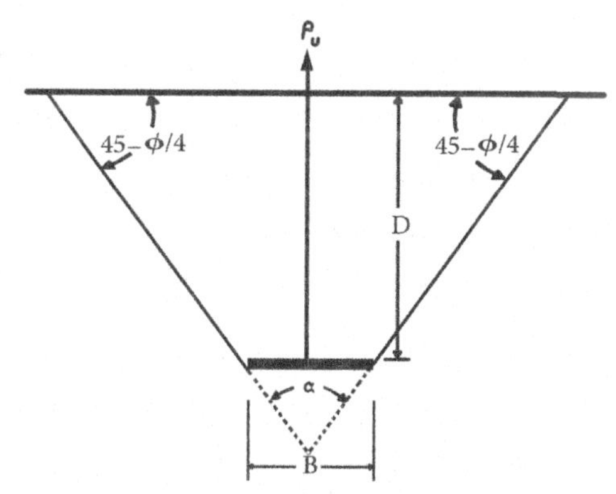

Figure 2.37 Failure surface assumed by Mors (1959).

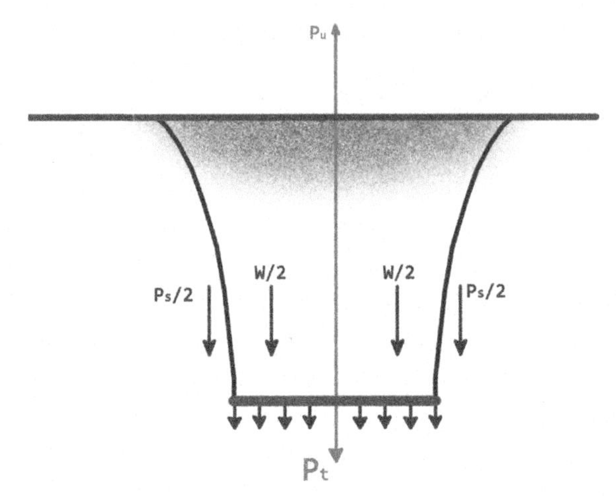

Figure 2.38 Anchor under pullout test by Balla (1961).

where, P_u, ultimate pullout force; w, effective weight of soil located in the failure zone; P_s, shearing resistance in failure zone; and P_t, force below the area.

In this case of sands P_t is equal to zero.

Balla (1961) proposed a method to predict the ultimate pullout capacity of the anchor plate. Ball developed a shearing resistance during the failure surface involved:

$$P_u = H^3\gamma\left[F_1\left(\varnothing, \frac{L}{D}\right) + F_3\left(\varnothing, \frac{L}{D}\right)\right] \tag{2.9}$$

The sum of F_1, F_3 can be obtained by Fig. 2.39. The breakout factor is defined as:

$$N_q = \frac{P_u}{\gamma AH} \tag{2.10}$$

Downs and Chieurzzi (1966), based on similar theoretical work, investigated whether the apex angle is always equal to 60 degrees, irrespective of the friction angle of the soil. Teng (1962) and Sutherland (1988) found that this assumption might be unsafe in many cases common with increase in depth. An approximate semi-empirical theory for the pullout loading force of horizontal strip, circular, and rectangular anchors has been proposed by Meyerhof and Adams (1968).

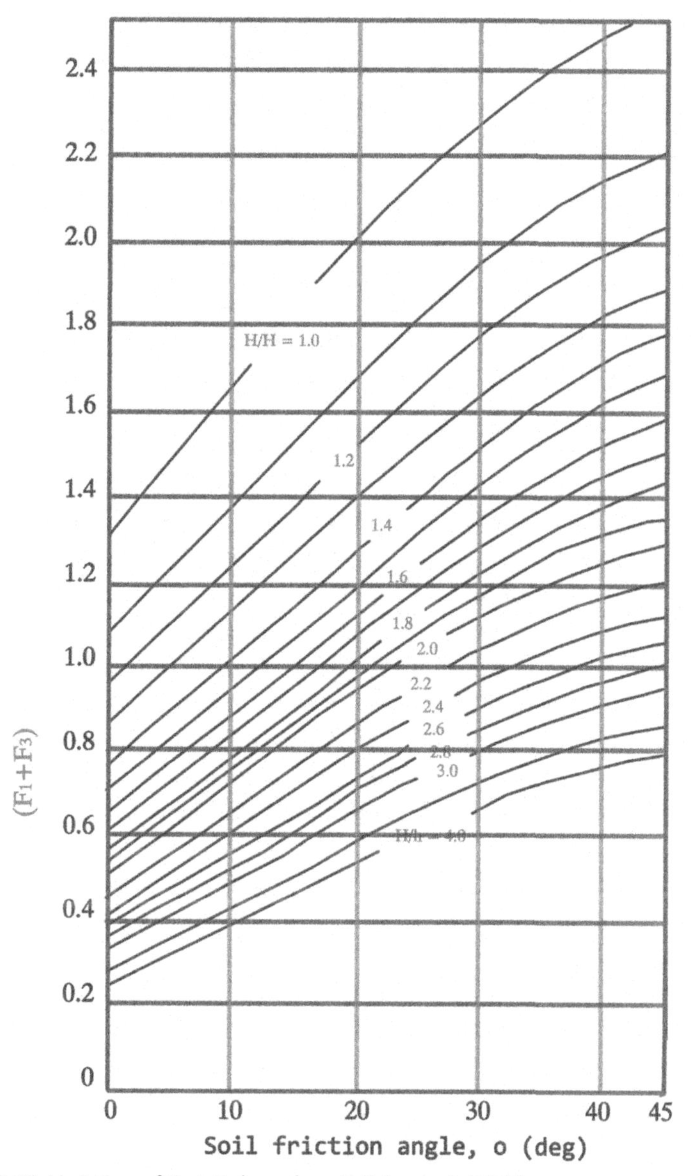

Figure 2.39 Variation of $F_1 + F_3$ based on Balla's result (1961).

For a strip anchor, an expression for the ultimate pullout capacity was selected by considering the equilibrium of the block of soil directly above the anchor (i.e., contained within the zone made when vertical planes are extended from the anchor edges). The capacity was assumed to act along the vertical planes extending from the anchor shape, while the total passive earth pressure was assumed to act at some angle to these vertical planes. This angle was selected based on laboratory test results while the passive earth pressures were evaluated from the results of Caquot and Kerisel (1948). For shallow plate anchors where the failure surface develops to the soil surface, the ultimate pullout capacity was determined by considering equilibrium of the material between the anchor and soil surface. For a deep anchor the equilibrium of a block of soil extending a vertical distance H above the anchor was presented, where H was less than the actual embedment depth of the plate anchor. The magnitude of H was determined from the observed extent of the failure surface from laboratory works.

The analysis of strip footings was developed by Meyerhof and Adams (1968) to include circular plate anchors using a semi-empirical shape factor to modify the passive earth pressure obtained for the plane strain case. The failure surface was assumed to be a vertical cylindrical surface through the anchor edge and extending to the soil surface. An approximate analysis for the capacity of rectangular plate anchors was selected as for downward loads (Meyerhof, 1951), by assuming the ground pressure along the circular perimeter of the two end portions of the failure surface was governed by the same shape factor assumed for circular anchors. It was, however, based on two key adoptions: namely, the edge of the failure surface and the distribution of stress along the failure surface. Even so, the theory presented by Meyerhof and Adams (1968) has been found to give reasonable estimates for a wide range of plate anchor problems. It is one of only two methods available for appraising the force of rectangular plate anchors.

Meyerhof and Adams (1968) expressed the ultimate pullout capacity in rectangular anchor plates as in the equation below (Figs. 2.40 and 2.41):

$$P_u = W + \gamma H^2(2S_f L + B - L)K_u \tan \varnothing \tag{2.11}$$

$$S_f = 1 + m\frac{L}{D}K_u = \text{Fig. 2.40} \tag{2.12}$$

$$N_q = 1 + \frac{L}{D}K_u \tan \varnothing \; m = \text{Fig. 2.41} \tag{2.13}$$

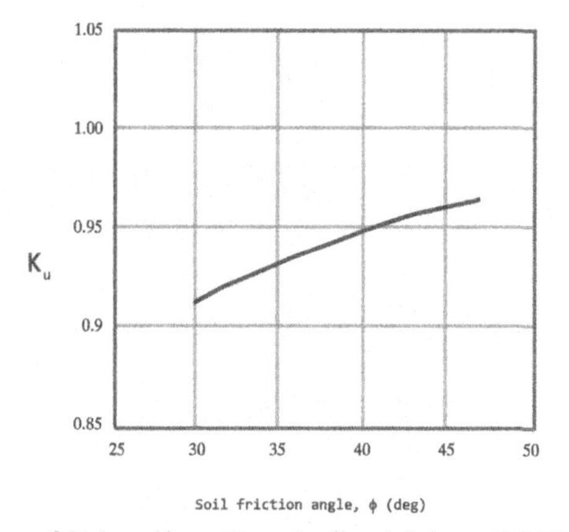

Figure 2.40 Variation of K_u based on Meyerhof and Adams (1968).

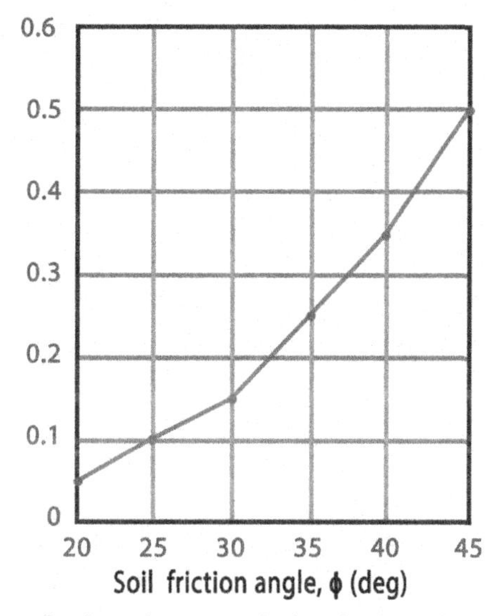

Figure 2.41 Variation of *m* based on Meyerhof and Adams (1968).

Vesic (1971) studied the problem of an explosive point charge expanding a spherical close to the surface of a semi-infinite, homogeneous and isotropic soils, as illustrated in Figs. 2.42 and 2.43.

$$P_u = \gamma H A N_q \tag{2.14}$$

$$N_q = \left[1 + A_1\left(\frac{H}{\frac{h_1}{2}}\right) + A_2\left(\frac{H}{\frac{h_1}{2}}\right)^2\right] \tag{2.15}$$

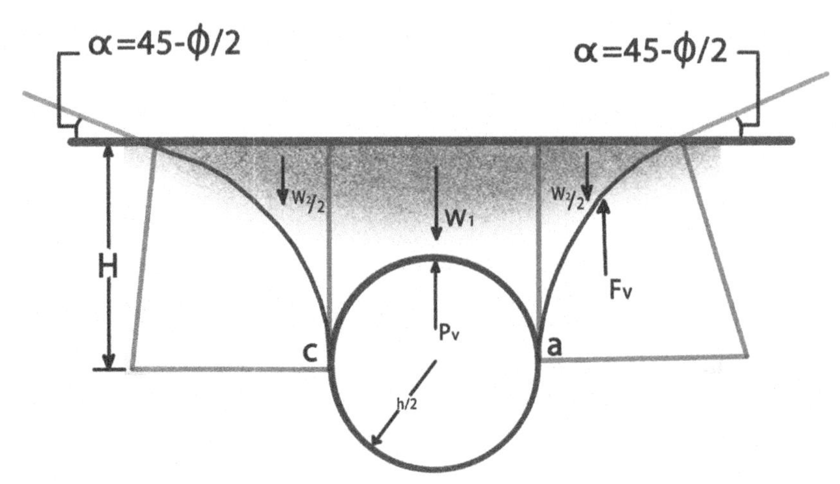

Figure 2.42 View of tests by Vesic (1971).

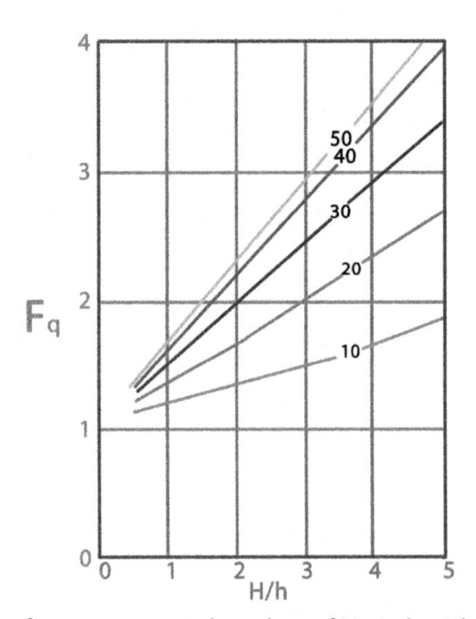

Figure 2.43 Breakout factor in strip anchor plate of Vesic (1971).

Clemence and Veesaert (1977) showed a formulation for shallow circular anchors in sand assuming a linear failure making an angle of $\beta = \varnothing/2$ with the vertical through the shape of the anchor plate as shown in Fig. 2.44. The contribution of shearing resistance along the length of

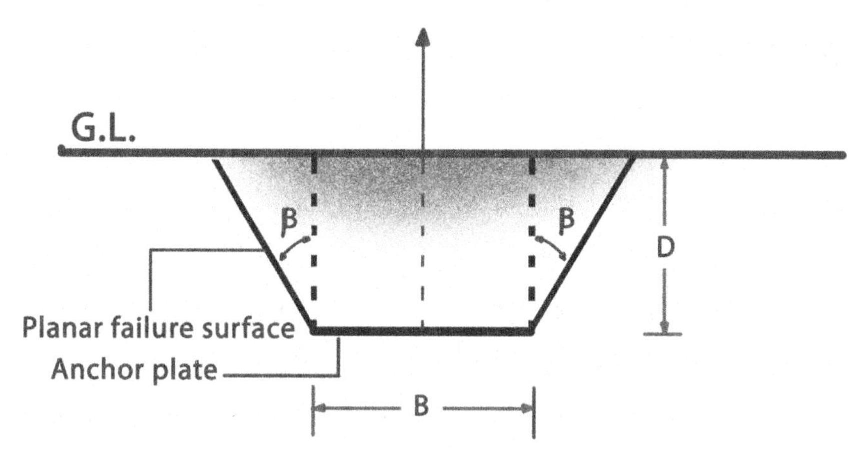

Figure 2.44 Failure surface assumed by Clemence and Veesaert (1977).

failure surface was approximately taken into consideration by selecting a suitable value of ground pressure coefficient from laboratory model works. The net ultimate pullout capacity can be given as:

$$P_u = \gamma V + \pi\gamma K_o \tan \varnothing \cos^2\left(\frac{\varnothing}{2}\right)\left(\frac{BD^2}{2} + \frac{D^3 \tan \frac{\varnothing}{2}}{3}\right) \tag{2.16}$$

where, V is the volume of the truncated cone above the anchor, and K_o is the coefficient of lateral earth pressure; they suggested that the magnitude of K_o may vary between 0.6 and 1.5 with an average value of about 1.

The finite element method had also been used by Vemeer and Sutjiadi (1985), Tagaya et al. (1983, 1988), and Sakai and Tanaka (1998). Unfortunately, only limited results were presented in these researches.

Rowe and Davis (1982) presented a research of the behavior of anchor plate in sand, as illustrated in Fig. 2.45. Tagaya et al. (1983, 1988) conducted two-dimensional plane strain and axi-symmetric finite element analyses using the constitutive law of Lade and Duncan (1975). Scale effects for circular plate anchors in dense sand were investigated by Sakai and Tanaka (1998) using a constitutive model for a non-associated strain hardening-softening elastoplastic material. The effect of shear band thickness was also introduced.

Koutsabeloulis and Griffiths (1989) investigated the trapdoor problem using the initial stress finite element method. Both plane strain and

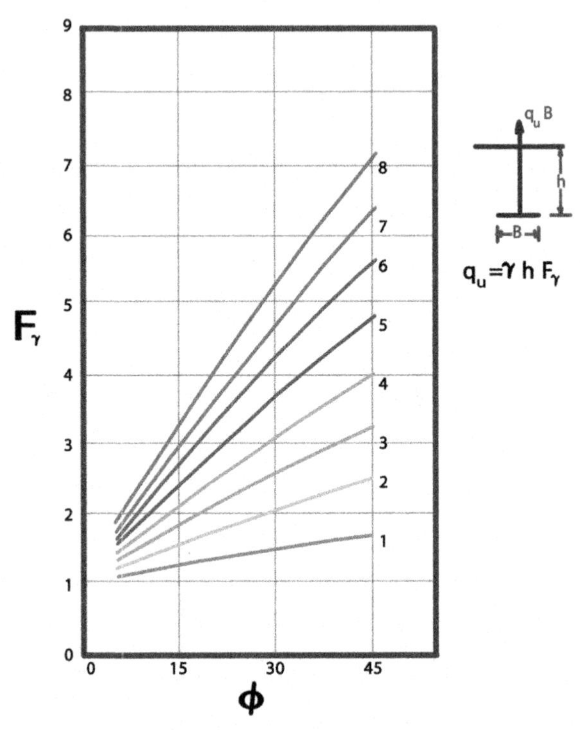

Figure 2.45 Variation of capacity factor F_γ in Rowe and Davis (1982).

axi-symmetric research were conducted. The researchers concluded that an associated flow rule has little effect on the collapse load for strip plate anchors but a significant effect (30%) on circular anchors. Large displacements were observed for circular plate anchors prior to collapse. In the LEM, an arbitrary failure surface is adopted along with a distribution of stress along the selected surface. Equilibrium conditions are then considered for the failing soil mass and an estimate of the collapse load is assumed. In the research of horizontal anchor force, the failure mechanism is generally assumed to be log spiral in edge (Ghaly & Hanna, 1994; Murray & Geddes, 1987; Saeedy, 1987; Sarac, 1989) and the distribution of stress is obtained by using either Kotter's equation (Balla, 1961), or by using an assumption regarding the orientation of the resultant force acting on the failure plane. The function of Murray and Geddes (1987) involves:

$$N_q = 1 + \frac{L}{D}\tan\varnothing\left(1 + \frac{D}{B} + \frac{\pi}{3}\frac{L}{B}\right) \tag{2.17}$$

Upper and lower bound limit analysis techniques have been used studies by Murray and Geddes (1987, 1989), Basudhar and Singh (1994) and Smith (1998) to estimate the capacity of horizontal and vertical strip plate anchors. Basudhar and Singh (1994) selected estimates using a generalized lower bound procedure based on finite elements and non-linear programming similar to that of Sloan (1988). The solutions of Murray and Geddes (1987, 1989) were selected by manually constructing cinematically admissible failure mechanisms (upper bound), while Smith (1998) showed a novel rigorous limiting stress field (lower bound) solution for the trapdoor problem.

Only a few investigations into the performance of ultimate pullout loading in sand were recorded in model numerical studies. One example of this is Fargic and Marovic (2003) which discussed the pullout capacity of plate anchors in soil under applied vertical force. Computation of the pullout and uplift force was performed by the finite element method. For a gravity load, the concept of initial stresses in Gauss points was selected. In the first increment of computation, these stresses were added to the vector of total stress. The soil was modeled by an elastoplastic constituent material model and the associated flow rule was used. The soil mechanics parameters of samples were determined by standard tests conducted on disturbed samples. For a complex constitutive numerical model of material to describe an actual state of soil, a greater number of soil mechanics parameters must be available. The tensile strength of the soil materials was crucial only in a few cases, and the problem of tensile plate anchors is one of them. An iterative procedure was used as the first procedure. The elements with tensile stresses were excluded from the following steps by diminishing the different modulus. More sophisticated constitutive laws are required for an exact analysis, and an adequate finite element method code program has to be prepared.

Merifield and Sloan (2006) used many numerical solutions for the analysis of plate anchors, as illustrated in Fig. 2.28. Until this time, very few rigorous numerical analyses had been performed to determine the pullout capacity of plate anchors in sand, as illustrated in Fig. 2.29. Although it is essential to verify theoretical solutions/numerical analysis with experimental studies wherever possible, results selected from their laboratory testing alone were typically problem specific, as illustrated in Fig. 2.30. It was particularly the case in geotechnicals, where they were dealing with a highly non-linear material that sometimes displays pronounced scale effects.

As the cost of performing laboratory works on each and every field problem combination is prohibitive, it is necessary to be able to model soil pullout loading numerically for the purposes of design. Existing numerical analyses generally assumed a condition of plane strain for the case of a continuous strip plate anchor or axi-symmetry for the case of circular plate anchors. The researchers were unaware of any three-dimensional numerical analyses to ascertain the effect of plate anchor shape on the uplift capacity.

Dickin and Laman (2007) investigated the numerical modeling of plate anchors by PLAXIS, which is finite element software. The numerical model was 0.57 m long, 0.46 m wide, and 0.23 m deep. Numerical analysis research investigated the uplift response of 1-m-wide strip anchors in sand and the results indicated maximum ultimate pullout capacity increase with anchor embedment ratio and sand packing. The research was carried out using a plane strain model for anchors in both loose and dense sand, as illustrated in Figs. 2.46 and 2.47. During the generation of the mesh, 15-node triangular elements were obtained in the determination of stresses.

Kumar and Bhoi (2008) used a group of multiple strip plate anchors placed in sand and subjected to equal magnitudes of vertical upward pullout loads which had been determined by numerical solutions, as illustrated in Fig. 2.33. Instead of using a number of anchor plates in numerical

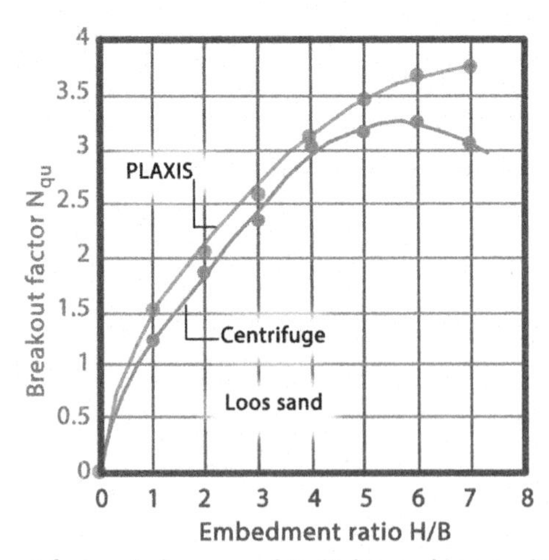

Figure 2.46 Breakout factors in loose sand in Dickin and Laman (2007).

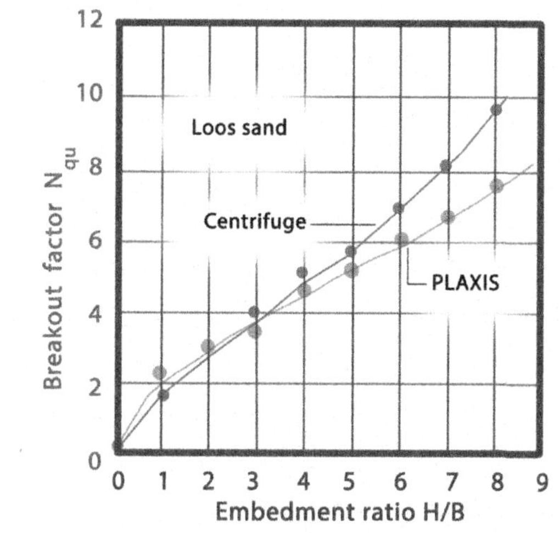

Figure 2.47 Breakout factors in dense sand in Dickin and Laman (2007).

modeling, a single plate anchor was used by modeling the boundary conditions along the plane of symmetry on both the sides of the plate anchor. The effect of interference due to a number of multiple strip plate anchors placed in a granular medium at different embedment depths was investigated by conducting a series of small numerical models.

Kuzer and Kumar (2009) used a group of two spaced strip plate anchors. The vertical pullout loading of two interfering rigid rough strip anchors embedded horizontally in sand, as illustrated in Fig. 2.34. The analysis was performed by obtaining an upper bound theorem of limit analysis combined with finite element and linear programing. The authors used an upper bound finite element limit analysis; the efficiency factor ξ_γ was computed for a group of two closely spaced strip plate anchors in sand. Previous numerical analyses on horizontal circular anchor plates in cohesionless soil can be found in Table 2.6.

2.3.1 Limitations of Numerical Work

Some previous researchers reported dealing with analysis of the limiting different numerical methods in ultimate pullout capacity. Very few rigorous numerical studies have been undertaken to determine anchor behavior. It is generally agreed that existing theories do not describe the behavior of anchor plates in enough detail (Sutherland, 1988). Most methods of analysis are based on the initial assumption of a common

Table 2.6 Previous numerical analyses on horizontal circular anchor plates in cohesionless soil

Researcher	Analysis method	Friction angle	*L/D*
Meyerhof and Adams (1968)	Limit equilibrium	–	–
Vesic (1971)	Cavity expansion	0–50°	0–5
Tagaya et al. (1988)	Elastoplastic Finite element	31.6°, 35.1°, 42°	0–30
Saeedy (1987)	Limit equilibrium	20–45°	1–10
Murray and Geddes (1987)	Limit analysis and limit equilibrium	All	All
Koutsabeloulis and Griffiths (1989)	Finite element method—initial stress	20°, 30°, 40°	1–8
Sarac (1989)	Limit equilibrium	0–50°	1–4
Ghaly and Hanna (1994)	Limit equilibrium	30–46°	1–10
Sakai and Tanaka (1998)	Elastoplastic Finite element	Dense	1–3
Merifield and Sloan (2006)	Limit analysis	20–40°	1–10

failure mode (LEM and upper bound limits analysis) in non-reinforced sand. Given that few attempts have been made to accurately monitor temporary soil deformations under laboratory conditions, the validity of the assumed failure mechanisms remain largely unproven in non-reinforced sand. Different symmetrical anchor plates and soil parameters were employed by different researchers. Inevitably such a wide range of parameters will contribute to the conflicting conclusions for ultimate uplift load of symmetrical anchor plates. Most anchor studies have been concerned with either vertical or horizontal uplift resistance in non-reinforced sand. However, anchors and researchers are frequently placed at orientations somewhere between horizontal and vertical depending on the type of application and load orientation (i.e., seawall, retaining walls, and transmission tower foundations). The effect of soil reinforcement on uplift response of symmetrical anchor plates needs to be investigated. The effects of soil reinforcement have largely been ignored. The effects of breakout factor and uplift response on symmetrical anchor plates in reinforced sand have largely been ignored. The effect that this may have on soil deflections needs to be investigated.

REFERENCES

Adams, J. I., & Hayes, D. C. (1967). The uplift capacity of shallow foundations. *Ontario Hydro-Research Quarterly, 19*, 1–13.

Andreadis, A., Harvey, R., & Burley, E. (1981). Embedded anchor response to uplift loading. *Journal of Geotechnical Engineering, 107*(1), 59–78.

Baker, W. H., & Konder, R. L. (1966). Pullout load capacity of a circular earth anchor buried in sand. *Highway Research Record, 108*, 1–10.

Balla, A. (1961). The resistance of breaking-out of mushroom foundations for pylons. *Proceedings of the Fifth International Conference on Soil Mechanics and Foundation Engineering, 1*, 569–576.

Baset, R. H. (1977). Underreamed ground anchors. *Proceedings of the Ninth International Conference on Soil Mechanics and Foundation Engineering, 1*, 11–17.

Basudhar, P. K., & Singh, D. N. (1994). A generalized procedure for predicting optimal lower bound break-out factors of strip anchors. *Geotechnique, 44*(2), 307–318.

Bouazza, A., & Finlay, T. W. (1990). Uplift capacity of plate anchors in a two-layered sand. *Geotechnique, 40*(2), 293–297.

Caquot, A. I., & Kerisel, J. (1948). Tables for the calculation of passive pressure, active pressure, and bearing capacity of foundations. In *Libraire du Bureau des Longitudes, de L'ecole Polytechnique*. Paris: Gauthier-Villars, Imprimeur-Editeur.

Clemence, S. P., & Veesaert, C. J. (1977). Dynamic pullout resistance of anchors in sand. In *International symposium on soil-structure interaction* (pp. 389–397), January 3 through January 7, 1977. Roorkee, India.

Das, B. M., & Seeley, G. R. (1975). Breakout resistance of shallow horizontal anchors. *Journal of Geotechnical Engineering, 101*(9), 999–1003.

Dickin, E. A. (1987). Uplift behaviour of horizontal anchor plates in sand. *Journal of the Soil Mechanics and Foundation Engineering Division, 114*(SM11), 1300–1317.

Dickin, E. A. (1988). Uplift behaviour of horizontal anchor plates in sand. *Journal of Geotechnical Engineering, 114*(11)), 1300–1317.

Dickin, E. A. (1994). Uplift resistance of buried pipelines in sand. *Soils and Foundations, 34*(2), 41–48.

Dickin, E. A., & Laman, M. (2007). Uplift response of strip anchors in cohesionless soil. *Journal of Advances in Engineering Software, 38*, 618–625.

Dickin, E. A., & Leung, C. F. (1983). Centrifuge model tests on vertical anchor plates. *Journal of Geotechnical Engineering, 109*(12), 1503–1525.

Downs, D. I., & Chieurzzi, R. (1966). Transmission tower foundations. *Journal of the Power Division, 88*(2), 91–114.

El Hansy, R. M. (1980). *Behaviour of shallow anchors* (MSc. thesis). Alexandria University, Alexandria, Egypt.

El Sawwaf, M. A. (2007). Uplift behavior of horizontal anchor plates buried in geosynthetic reinforced slopes. *Geotechnical Testing Journal, 30*(5), 418–426.

Fadl, M. O. (1981). *The behaviour of plate anchors in sand* (PhD. thesis). University of Glasgow, Glasgow, Scotland.

Fargic, L., & Marovic, P. (2003). Pullout capacity of spatial anchors. *Engineering Computations, 21*(6), 598–700.

Frydman, S., & Shamam, I. (1989). Pullout capacity of slab anchors in sand. *Canadian Geotechnical Journal, 26*, 385–400.

Ghaly, A. M. (1997a). Load displacement prediction for horizontally loaded vertical plates. *Journal of Geotechnical and Geoenvironmental Engineering, 123*(1), 74–76.

Ghaly, A. M. (1997b). The static equilibrium of drag anchors in sand: discussion. *Canadian Geotechnical Journal, 34*(4), 635–636.

Ghaly, A. M., & Hanna, A. M. (1994). Ultimate pullout resistance of single vertical anchors. *Canadian Geotechnical Journal, 31*(5), 661–672.

Ghaly, A. M., Hanna, A. M., & Hanna, M. (1991). Uplift behavior of screw anchors in sand. I: Dry sand. *Journal of Geotechnical Engineering, 117*(5), 773–793.

Giffels, W. C., Graham, R. E., & Mook, J. F. (1960). Concrete cylinder anchors. *Electrical World, 154*, 46–49.

Hanna, T. H., & Carr, R. W. (1971). The loading behaviour of plate anchors in normally and over consolidated sands. In *Proceedings of the fourth international conference on soil mechanics and foundation engineering* (pp. 589–600). Budapest, Hungary.

Hanna, T. H., Sparks, R., & Yilmaz, M. (1971). Anchor behaviour in sand. *Journal of the Soil Mechanics and Foundation Engineering Division ASCE, 98*(11), 1187–1208.

Hoyt, R. M., & Clemence, S. P. (1989). Uplift capacity of helical anchors in soil. In *Proceedings of the Regional South America conference on soil mechanics and foundation engineering* (Vol. 2, pp. 1019–1022). Rio de Janeiro, Brazil.

Ilamparuthi, K., & Dickin, E. A. (2001). The influence of soil reinforcement on the uplift behaviour of belled piles embedded in sand bed. *Geotextiles and Geomembranes, 19*, 1–22.

Ilamparuthi, K., Dickin, E. A., & Muthukrisnaiah, K. (2002). Experimental investigation of the uplift behaviour of circular plate anchors embedded in sand. *Canadian Geotechnical Journal, 39*(3), 648–664.

Ilamparuthi, K., & Muthukrisnaiah, K. (1999). Anchors in sand bed: Delineation of rupture surface. *Ocean Engineering, 26*, 1249–1273.

Ilamparuthi, K., & Muthukrishnaiah, K. (2001). Breakout capacity of seabed anchors due to snap loading. In *Proceedings of international conference in ocean engineering* (pp. 393–400). IIT Madras, Chennai, India.

Ilamparuthi, K., Ravichandran, P., & Mohammed Toufeeq, M. (2008). Study on uplift behaviour of plate anchor in Geogrid reinforced sand bed. In *Geotechnical earthquake engineering and soil dynamics IV 2008. Geotechnical special publication series no 181* (pp. 1190–1199), May 18–22, 2008, Sacramento, CA. Reston, VA: American Society of Civil Engineers.

Ireland, H. O. (1963). Uplift resistance of transmission tower foundations. *Journal of the Power Division ASCE, 89*(PO1), 115–118.

Johnston, R. S. (1984). *Pull-out testing of Tensar Geogrids* (Masters thesis). University of California, Davis, CA, USA, 179pp.

Kanakapura, S., Rao, S., & Kumar, J. (1994). Vertical uplift capacity of horizontal anchors. *Journal of Geotechnical Engineering ASCE, 120*(7), 1134–1147.

Kananyan, A. S. (1966). Experimental investigation of the stability of bases of anchor foundations. *Osnovanlya, Fundamenty i mekhanik Gruntov, 4*(6), 387–392.

Kingshri, A., Ilamparuthi, K., & Ravichandran, P. T. (2005). Enhancement of uplift capacity of anchors with Geocomposite. In *Proceeding of national symposium on geotechnical prediction methods (Geopredict 2005)* (pp. 148–152). IIT Madras, Chennai, India.

Koutsabeloulis, N. C., & Griffiths, D. V. (1989). Numerical modelling of the trap door problem. *Geotechnique, 9*(1), 77–89.

Krishna, Y. S. R. (2000). *Numerical analysis of large size horizontal strip anchors* (PhD. thesis). Indian Institute of Science, Bengaluru, India.

Krishnaswamy, N. R., & Parashar, S. P. (1992). Effect of submergence on the uplift resistance of footings with geosynthetic inclusion. In *Proceedings of Indian geotechnical conference* (pp. 333–336). Surat, India.

Krishnaswamy, N. R., & Parashar, S. P. (1994). Uplift behaviour of plate anchors with Geosynthetics. *Geotextiles and Geomembranes, 13*, 67–89.

Kumar, J., & Bhoi, M. K. (2008). Interference of multiple strip footings on sand using small scale model tests. *Geotechnical and Geological Engineering, 26*(4), 469–477.

Kuzer, K. M., & Kumar, J. (2009). Vertical uplift capacity of two interfering horizontal anchors in sand using an upper bound limit analysis. *Computers and Geotechnics, 36*(6), 1084–1089.

Kwasniewski, J., Sulikowska, I., & Walker, A. (1975). Anchors with vertical tie rods. In *Proceedings of the first Baltic conference, soil mechanics and foundation engineering* (pp. 122–133). Gdansk, Poland.

Lade, P. V., & Duncan, J. M. (1975). Elasto-plastic stress–strain theory for cohesionless soil. *Journal of the Soil Mechanics and Foundation Engineering Division ASCE, 101*(10), 1037–1053.

Liu, H., Li, Y., Yang, H., Zhang, W., & Liu, C. (2010). Analytical study on the ultimate embedment depth of drag anchors. *Ocean Engineering, 37*(14–15), 1292–1306.

Liu, M., Liu, J., & Gao, H. (2010). Displacement field of an uplifting anchor in sand. *Deep Foundations and Geotechnical In Situ Testing*, Geotechnical Special Publication (Issue 205 GSP), 261–266.

Majer, J. (1955). Zur berechnung von zugfundamenten. *Osterreichister Bauzeitschift, 10*(5), 85–90, (in German).

Mariupolskii, L. G. (1965). The bearing capacity of anchor foundations. *SMFE, Osnovanlya, Fundamenty i mekhanik Gruntov, 3*(1), 14–18.

Merifield, R., & Sloan, S. W. (2006). The ultimate pullout capacity of anchors in frictional soils. *Canadian Geotechnical Journal, 43*(8), 852–868.

Meyerhof, G. G. (1951). Ultimate bearing capacity of footings on sand layer overlaying clay. *Canadian Geotechnical Journal, 11*(2), 223–229.

Meyerhof, G. G., & Adams, J. I. (1968). The ultimate uplift capacity of foundations. *Canadian Geotechnical Journal, 5*(4), 225–244.

Mitsch, M. P., & Clemence, S. P. (1985). The uplift capacity of helix anchors in sand. In *Proceedings of a session sponsored by the Geotechnical Engineering Division of the ASCE* (pp. 26–47). Detroit, MI, USA.

Mors, H. (1959). The behaviour of most foundations subjected to tensile forces. *Bautechnik, 36*(10), 367–378.

Murray, E. J., & Geddes, J. D. (1987). Uplift of anchor plates in sand. *Journal of Geotechnical Engineering, 113*(3), 202–214.

Murray, E. J., & Geddes, J. D. (1989). Resistance of passive inclined anchors in cohesionless medium. *Géotechnique, 39*(3), 417–431.

Murray, E. J., & Geddes, J. D. (1996). Plate anchor groups pulled vertically in sand. *Journal of Geotechnical Engineering, 122*(7), 509–516.

Neely, W. J., Stuart, J. G., & Graham, J. (1973). Failure loads of vertical anchor plates in sand. *Journal of the Soil Mechanics and Foundation Engineering Division ASCE, 99*(SM9), 669–685.

Ovesen, N. K. (1981). Centrifuge tests on the uplift capacity of anchors. In *Proceedings of the 10th international conference on soil mechanics and foundation engineering* (pp. 717–722). Stockholm, Sweden.

Pearce, A. (2000). *Experimental investigation into the pullout capacity of plate anchors in sand* (MSc. thesis). University of Newcastle, Newcastle, NSW, Australia.

Rajagopal, K., & SriHari, V. (1996). Analysis of anchored retaining walls. In H. Ochiai, N. Yaufuku, & K. Omine (Eds.), *Earth reinforcement: Proceedings of the international symposium on earth reinforcement, Fukuoka, Kyushu, Japan, 1996, Vol. 1* (pp. 475–479). Rotterdam and Brookfield, VT: A.A. Balkema.

Ramesh Babu, R. (1998). *Uplift capacity and behaviour of shallow horizontal anchors in soil* (PhD. thesis). Indian Institute of Science, Bengaluru, India.

Ravichanadran, P. T., & Ilamparuthi, K. (2004). Behaviour of rectangular plate anchors in reinforced and unreinforced sand beds. In *Proceedings of ICCGE* (pp. 123–128). Mumbai, India.

Rowe, R. K. (1978). *Soil structure interaction analysis and its application to the prediction of anchor behaviour* (PhD. thesis). University of Sydney, Sydney, NSW, Australia.

Rowe, R. K., & Davis, E. H. (1982). The behaviour of anchor plates in sand. *Geotechnique, 32*(1), 25–41.

Saeedy, H. S. (1987). Stability of circular vertical anchors. *Canadian Geotechnical Journal, 24*, 452–456.
Sakai, T., & Tanaka, T. (1998). Scale effect of a shallow circular anchor in dense sand. *Soils and Foundations, Japan, 38*(2), 93–99.
Sarac, D. Z. (1989). Uplift capacity of shallow buried anchor slabs. *Proceedings of the 12th International Conference on Soil Mechanics and Foundation Engineering, 12*(2), 1213–1218.
Saran, S., Ranjan, G., & Nene, A. S. (1986). Soil anchors and constitutive laws. *Journal of Geotechnical Engineering, 112*(12), 1084–1100.
Selvadurai, A. P. S. (1989). The enhancement of the uplift capacity of buried pipelines by the use of geogrids. *Geotechnical Testing Journal ASTM, 12*, 211–216.
Selvadurai, A. P. S. (1993). Uplift behaviour of strata grid anchored pipelines embedded in granular soils. *Geotechnical Engineering, 24*, 39–55.
Sergeev, I. T., & Savchenko, F. M. (1972). Experimental investigations of soil pressure on the surface of an anchor plate. *Journal of the Soil Mechanics and Foundation Engineering, 9*(5), 298–300.
Singh, R. B. (1992). Anchored earth technique using semi-Z shaped mild steel anchors. In H. Ochiai, S. Hayashi, & K. Otani (Eds.), *Earth reinforcement practice: Proceedings of the international symposium on earth reinforcement, Fukuoka, Kyushu, Japan, 1992, Vol. 1* (pp. 419–424). Rotterdam and Brookfield, VT: A.A. Balkema.
Sloan, S. W. (1988). Lower bound limit analysis using finite elements and linear programming. *International Journal for Numerical and Analytical Methods in Geomechanics, 12*(1), 61–67.
Smith, C. C. (1998). Limit loads for an anchor/trapdoor embedded in an associated coulomb soil. *International Journal for Numerical and Analytical Methods in Geomechanics, 22* (11), 855–865.
Subbarao, C., Mukhopadhyay, S., & Sinha, J. (1988). Geotextile ties to improve uplift resistance of anchors. In *Proceedings of the first Indian geotextile conference on reinforced soil and geotextiles* (pp. F3–F8). Mumbai, India.
Sutherland, H. B. (1965). Model studies for shaft raising through cohesionless soils. *Proceedings of the Sixth International Conference on Soil Mechanics and Foundation Engineering, 2*, 410–413.
Sutherland, H. B. (1988). Uplift resistance of soils. *Geotechnique, 38*(4), 493–516.
Tagaya, K., Scott, R. F., & Aboshi, H. (1988). Pullout resistance of buried anchor in sand. *Soils and Foundations, Japan, 28*(3), 114–130.
Tagaya, K., Tanaka, A., & Aboshi, H. (1983). Application of finite element method to pullout resistance of buried anchor. *Soils and Foundations, Japan, 23*(3), 91–104.
Teng, W. C. (1962). *Foundation design*. Englewood Cliffs, NJ: Prentice-Hall.
Turner, E. Z. (1962). Uplift resistance of transmission tower footings. *Journal of the Power Division ASCE, , 88*(PO2), 17–33.
Udwari, J. J., Rodgers, T. E., Jr., & Singh, H. (1979). A rational approach to the design of high capacity multi-helix screw anchors. In *Proceedings of the seventh IEEE/PES, transmission and distribution conference and exposition* (pp. 606–610). New York, NY, USA.
Vermeer, P. A., & Sutjiadi, W. (1985). The uplift resistance of shallow embedded anchors. *Proceedings of the 11th International Conference on Soil Mechanics and Foundation Engineering, 4*, 1635–1638.
Vesic, A. S. (1971). Breakout resistance of objects embedded in ocean bottom. *Journal of the Soil Mechanics and Foundation Engineering Division ASCE, 97*(9), 1183–1205.
Vesic, A. S. (1972). Expansion of cavities in infinite soil mass. *Journal of the Soil Mechanics and Foundation Engineering Division ASCE, 98*(3), 265–290.
Wang, M. C., & Wu, A. H. (1980). Yielding load of anchor in sand. In *Proceedings of the ASCE, application of plasticity and generalized stress-strain in geotechnical engineering* (pp. 291–307). Hollywood, FL, USA.

CHAPTER 3

Research Methodology

3.1 INTRODUCTION

The main objective of this research was to analyze and investigate the influence of various parameters such as reinforcement materials, soil types, and grid-fixed reinforcement (GFR) on the uplift capacity of symmetrical anchor plates of different shapes, such as circular, square, and rectangular. This could only be done when the cohesionless soil unit weight was controlled as the constant factor, thus implying the significance of the cohesionless soil placement method during the preparation prior to testing. The influence of anchor plate size, diameter, and shape on uplift capacity in non-reinforced, reinforced, and GFR-reinforced cohesionless soil was investigated. Various symmetrical anchor plates were used during the tests and numerical simulation was performed using PLAXIS in the course of this evaluation. Full details on the uplift response of anchor plates, comprising the placement method, preparatory steps, test equipments, and numerical simulations, are discussed in this chapter.

3.2 METHODS

Cohesionless soil packing is particularly important in uplift tests. Similar cohesionless soil unit weights were obtained as a basis for comparing the influence of uplift parameters on the symmetrical anchor plate capacity. A sand unit weight at a value of 15 kN/m^3 was defined for loose packing, whereas dense packing was defined at a unit weight of 17 kN/m^3.

3.2.1 Method Used to Obtain Loose Cohesionless Soil

Loose conditions were obtained using the cohesionless soil raining method. Trial tests were run in order to predict the particular conditions that had to exist before the target unit weight could be achieved.

Soil Reinforcement for Anchor Plates and Uplift Response.
DOI: http://dx.doi.org/10.1016/B978-0-12-809558-4.00003-0
© 2017 Elsevier Inc.
All rights reserved.

For cohesionless soil in loose conditions, trial tests indicated a limitation for sand layer thickness of 50 mm. Regarding the sand raining test, a range of falling cohesionless soil heights was employed in order to obtain the height required for the desired unit weight. The tests showed that a falling height of 450 mm for fine sand had to be maintained for every 50-mm layer to achieve an average dry unit weight of 15 kN/m^3.

3.2.2 Method Used to Obtain Dense Cohesionless Soil

Similar to loose cohesionless soil conditions, the standard thickness of the sand sample was taken as 50 mm. Results show that sand required a compaction time of 2 minutes per 50-mm layer to give an average dry unit weight value of 17 kN/m^3.

3.3 TEST EQUIPMENT IN THE SOIL LABORATORY

Test equipments and accessories required were fabricated for the uplift test.

3.3.1 Test Box

Uplift tests were carried out in a test box in the soil laboratory. The dimensions for the test box were 1000 mm × 500 mm × 1200 mm. Figs. 3.1 and 3.2 show the test box used for the uplift tests. The loading frames were also designed to suit the requirements of the tests.

3.3.2 Load Cell

A Japan-made, CLP-100 kA type load cell with a capacity of 1000 kgf was used during the tests. This load cell had a sensitivity of 1.5 mV/V and its coefficient was 3.24.

3.3.3 Linear Variable Displacement Transducer

The coefficient obtained for the linear variable displacement transducer (LVDT) was 0.005 using the direct calibration method with a maximum travel of 100 mm.

Figure 3.1 Experimental setup for uplift test on an anchor plate in sand.

3.3.4 Mechanical Compactor

The mechanical compactor was used to produce dense packing of the sample for the uplift tests.

3.3.5 Sand Rainer

Loose-packed cohesionless soil was obtained using the raining method. The diameter of the openings for cohesionless soil diffusion was approximately 1.0–1.2 mm with spacing between openings at an estimated distance of 1.0 mm.

3.3.6 Portable Datalogger

Readings obtained by load cell and LVDT were recorded by the datalogger and printed out every minute. The datalogger was a model TDS-301, and is illustrated in Fig. 3.3.

Figure 3.2 The uplift test setup in the laboratory.

Figure 3.3 Datalogger used for recording the experimental results.

Figure 3.4 Motor winch used for recording the experimental results.

3.3.7 Motor Winch

The motor winch controls the speed of uplift during the test. The speed was pre-set at approximately 0.1 mm/min to ensure a quasi-static rate of uplift test. Oriental Motor in Japan manufactured the motor with a maximum output of 40 W at 200 V running at 50 Hz. The motor winch is illustrated in Fig. 3.4.

3.3.8 Symmetrical Anchor Plates

The symmetrical anchor plates used were square, circular, and rectangular in shape. The plate dimensions included were 5 cm × 5 cm, 7.5 cm × 7.5 cm, and 10 cm × 10 cm for the square shapes, 20 cm × 5 cm and 30 cm × 7.5 cm for the rectangular shapes. The diameters of the circular plates were 5, 7.5, and 10 cm. Fig. 3.5 shows the various types of symmetrical anchor plates that were used in the uplift tests.

3.4 UPLIFT TEST

The uplift tests were conducted in the geotechnical laboratory. The main observations during the experimental tests were the stress–displacement

Figure 3.5 Symmetrical anchor plates.

relationships during symmetrical anchor plate breakout. The steps involved in the uplift test are described in the following sections.

3.4.1 Experimental Setup

Figs. 3.6 and 3.7 show the requirements for the uplift tests and the available equipments. The test box was made from a clear perspex sheet to show the failure patterns. The symmetrical anchor plate models were connected to a pulling tendon cable for uplifting. A quasi-static rate of pullout of approximately 1.5 mm/min was used for every test. This was to ensure that the symmetrical anchor plates surrounding the element would have ample time to redistribute during the uplift. Uplift capacity was measured by a load cell attached to the pulling tendon cable during uplift test. An LVDT was placed at the top of the symmetrical anchor plate holder to measure the vertical displacement, so as to record the amount of symmetrical anchor plate movement required to mobilize the ultimate uplift capacity. A motor was connected to the pulling tendon cable via steel tendon cables. A datalogger was used to record data from the load cell and the LVDT.

3.4.2 Uplift Test Procedures

The uplift test recorded the uplift capacity of the symmetrical anchor plates. This meant that only the symmetrical anchor plates were included

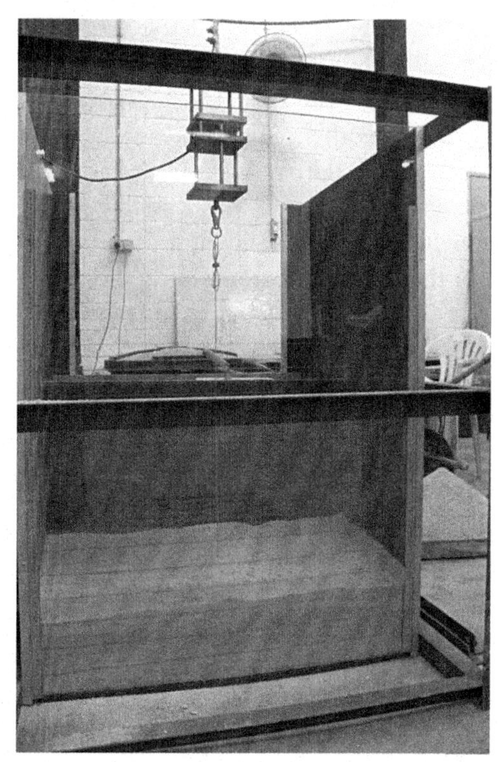

Figure 3.6 The setup of the uplift test.

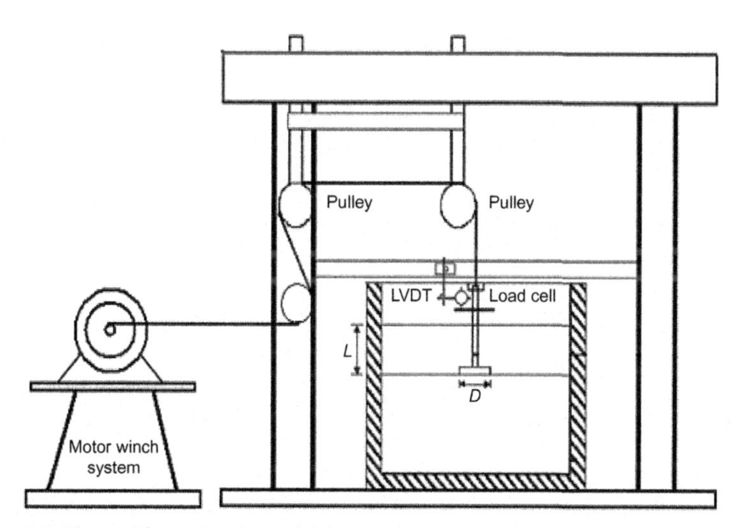

Figure 3.7 The uplift test in the soil laboratory.

in the analysis of uplift capacity. The uplift test procedures for symmetrical anchor plates included the following steps:

1. The symmetrical anchor plate model to be tested was attached to the tendon cable which was then connected to the load cell holder. All apparatus included in the test were controlled for symmetrical anchor plates in the test box.

   These controls included the following:

   a. inspection of the test frame to ensure rigidity;
   b. inspection of the pulling tendon cable to ensure that it had not worn out;
   c. checking that the test box was empty and free from cohesionless soil particles;
   d. checking that the tendon cable connected to the load cell holder was firmly in place.
2. The symmetrical anchor plate model to be tested was lowered slowly into the test box at the intended location marked beforehand to ensure that the vertical pullout was axially loaded. The symmetrical anchor plate was controlled again vertically using the spirit level.
3. Cohesionless soil was then placed in the box according to the placement method described earlier.
4. After the required height was reached, the surface layer was deformed, and the load cell and LVDT were then placed in position.
5. The calibration of the load cell and LVDT had been done earlier such that only the measurements of net uplift response and vertical displacement were fed into the datalogger.
6. The datalogger was then started to take readings at certain intervals.
7. It was noticed that the symmetrical anchor plate underwent failure when the peak value for the uplift response was reached.

Tests had to be repeated when the desired results were not achieved; causes included disturbance to any part of the experimental setup during testing, human error, and power shortage that resulted in the discontinuity of the test. All factors should have indicated that the data obtained from the tests were reliable before they were accepted for analysis.

3.5 TEST PROCEDURE

For the current research, the uplift test was conducted on sand and symmetrical anchor plates. Different test series were conducted on different soil types to research the effects of size, diameter, embedment ratio, dry unit weight, and shape of symmetrical anchor plates as follows: (1) test series

1–4 in non-reinforced cohesionless soil; (2) test series 5–8 in reinforced cohesionless soil; (3) test series 9 in non-reinforced, reinforced, and GFR cohesionless soil based on the finite element method software, PLAXIS.

3.5.1 Test Series 1: Influence of Symmetrical Anchor Plate Size on the Uplift Capacity in Non-reinforced Cohesionless Soil

A total of 32 tests were conducted to research the influence of symmetrical anchor plate sizes. The plates size were 50, 75, and 100 mm in square and circular anchor plates, whereas the rectangular plates were 50 and 75 mm in width and 200 and 300 mm in length for embedment ratios of 1–4 in non-reinforced sand.

3.5.2 Test Series 2: Influence of Symmetrical Anchor Plate Shapes on the Uplift Capacity in Non-reinforced Cohesionless Soil

Twelve uplift tests conducted to research the influence of symmetrical anchor plate shapes were performed on the symmetrical anchor plates with embedment ratios of 1–4. The symmetrical anchor plates used were circular, square, and rectangular anchor plates in non-reinforced cohesionless soil. The embedment ratios were chosen up to 4 since higher values would require smaller sizes. This would result in a stability problem since the anchor-plate–tendon-cable–load cell assemblage would topple due to the small symmetrical anchor plate base.

3.5.3 Test Series 3: Influence of Sand Unit Weight on the Uplift Capacity in Non-reinforced Cohesionless Soil

Sixty-four tests were performed in loose and dense sand. Sand unit weights of 15 and 17 kN/m^3 were used to compare the influence of sand packing on uplift. Test series 1–4 were conducted in loose and dense states in non-reinforced cohesionless soil. These values of unit weights were chosen since they were within the range of unit weights obtainable through laboratory means for the sand samples used. The range of sand unit weights tested is also typical for most sand types in loose and dense packing.

3.5.4 Test Series 4: Influence of Anchor Plate Embedded Ratio on the Uplift Capacity in Non-reinforced Cohesionless Soil

The discussion concerning embedment ratio involved 64 tests in total. The symmetrical anchor plates were used in the four embedded ratios (1–4) in non-reinforced cohesionless soils.

3.5.5 Test Series 5: Influence of Number and Vertical Spacing of Geogrid Layers on the Uplift Capacity

In this test series, geogrid layers were placed at equal vertical spacings of 0.5*D*, 0.75*D*, and 1*D* with the first layer resting between 0*D* and 0.5*D*. The geogrid layers were covered with soil after putting them in place. In practice or in laboratory conditions, geogrid layers can be used in a roll shape and placed in a specified area.

3.5.6 Test Series 6: Influence of Geogrid Layer Proximity to the Anchor on the Uplift Capacity

The symmetrical anchor plates were placed in loosely and densely packed sand with the inclusion of one geogrid layer using various distances of 0*D*, 0.5*D*, 0.75*D*, and 1*D* with $B/D = 1-2$ over the symmetrical anchor plates. Fig. 3.8 shows the schematic of geogrid layers in the uplift test. The different parameters include the number of geogrid layers (*N*), type of geogrid layers, vertical spacing between the layers (*u*), geogrid width (*B*), depth of the soil above the anchor plate (*L*), and the vertical spacing between the anchor plates and the geogrid layers (*x*).

3.5.7 Test Series 7: Influence of Number and Vertical Spacing of GFR Layers on the Uplift Capacity

GFR is a new system for tying the geosynthetic material into the ground. GFR is made of a fiber-reinforced polymer (FRP) strip and an end ball

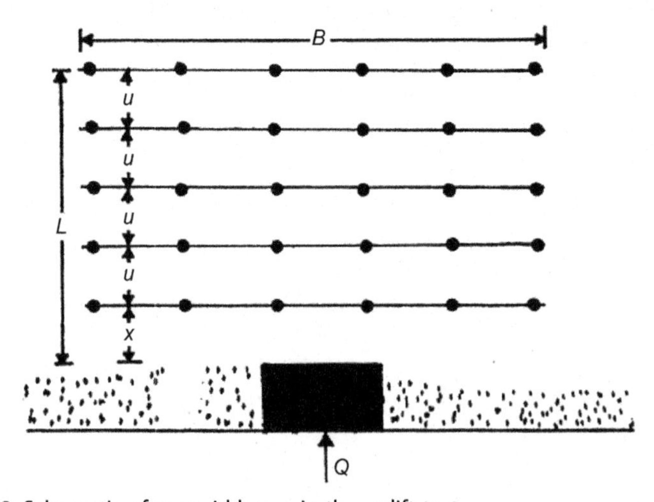

Figure 3.8 Schematic of geogrid layers in the uplift test.

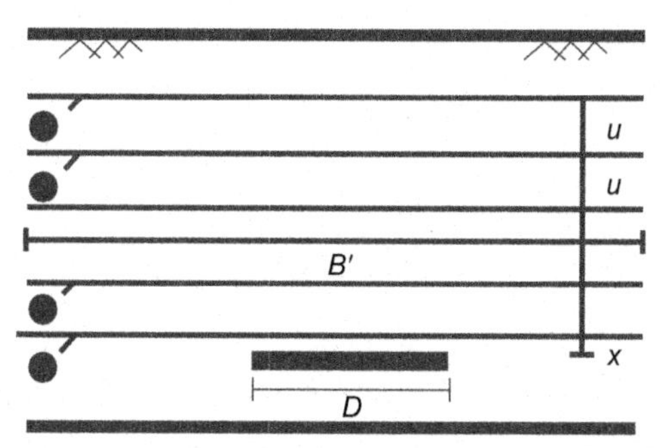

Figure 3.9 The schematic of GFR layers in the uplift test.

that ties up to the geosynthetic material and anchors it into the soil. In this test series, GFR layers were placed at equal vertical spacings of 0.5*D*, 0.75*D*, and 1*D*, with the first layer resting on the plate or at 0.5*D*. The GFR was made 10-mm diameter layers tied together and attached to the geogrid layer at an angle of 45 degrees. After this was done, the GFR layers and the geogrid layer were put in place and covered with soil, as illustrated in Fig. 3.9. In practice, the GFR layers can be assembled beforehand and then put in the specified place. However, for this purpose, GFR layers in the shape of a roll cannot be used.

3.5.8 Test Series 8: Influence of GFR Layer Proximity to the Anchor on the Uplift Capacity

A total of 30 tests were conducted to research the influence of symmetrical anchor plate sizes. One GFR layer was located at various distances of 0, 0.5*D*, 0.75*D*, and 1*D* above the symmetrical anchor plate in the tests. Fig. 3.9 shows the schematic of GFR layers in the uplift tests. The varied conditions include the vertical spacing between GFR layers (*u*), GFR width (*B'*), geogrid width (*B*), plate width (*D*), depth of soil above anchor plate (*L*) and vertical spacing between anchor plate and layer of GFR (*x*).

3.5.9 Test Series 9: Influence of Anchor Plate Size and Shape, and Sand Unit Weight on the Uplift Capacity in Non-reinforced, Reinforced, and GFR Cohesionless Soil Based on Numerical Simulation Using PLAXIS

This test series was conducted using numerical simulation of symmetrical anchor plates based on various sizes, shapes, and sand unit weights using PLAXIS in non-reinforced, reinforced, and GFR-reinforced cohesionless soil.

Table 3.1 Summary of uplift tests and simulations conducted for the various combinations of parameters

Parameter	Condition	Cohesionless soil type
Anchor plate sizes	50, 75, 100, 200, 300 mm	Non-reinforced
Anchor plate shapes	Circular, square, rectangular	Non-reinforced
Sand unit weight	Loose and dense	Non-reinforced
Anchor plate's embedded ratio	1, 2, 3, and 4	Non-reinforced
Number and vertical spacing of geogrid layers	$B/D=1-2$, $x/D=0-0.5$, $u/D=0-1$	Reinforced
Geogrid layer proximity to the anchor	$B/D=1-2$, $x/D=0-0.5$	Reinforced
Number and vertical spacing of GFR layers	$B'/D=1-2$, $x/D=0-0.5$, $u/D=0-1$	GFR reinforced
GFR layer proximity to the anchor	$B'/D=1-2$, $x/D=0-0.5$	GFR reinforced
Anchor plate sizes, shapes, and sand unit weight	Loose and dense, 50, 75, 100, 200, 300 mm	Non-reinforced, reinforced, GFR reinforced

3.6 SUMMARY OF EXPERIMENTAL AND NUMERICAL STEPS

A summary of the experimental and numerical steps divided into the nine test series is shown in Table 3.1.

The test procedure was considered adequate to cover the range of parameters under research and to systematically isolate the effects of a certain parameter on the uplift capacity. This enabled critical assessment of the experiments and the numerical simulations conducted, and provided a basis for comparison.

3.7 SOIL PROPERTIES

Several tests were performed to determine the properties of sand samples during experimental work. The tests included were:

1. Particle size distribution using the dry sieve method (BS 1377: Part 2:1990).

2. Maximum and minimum unit weight using the vibratory table method (ASTM standards on soil compaction, 1993 edition, Test designation D4254−91 and D4253−93).
3. Direct shear test using the small shear box (BS 1377: Part 7:1990).
4. Particle density using the small pyknometer method (BS 1377: Part 2: 1990).

3.7.1 Particle Size Distribution

The particle size distribution test was performed according to BS 1377: Part 2:1990 using the dry sieve method. This method covers the quantitative determination of the particle size distribution in a cohesionless soil down to the fine-sand size. For sand, three dry sieve tests were performed with sieve sizes as follows:

Aperture size: 2.36, 1.18, 0.6, 0.3, 0.212, 0.15, and 0.075 mm.

The sieve sizes used were considered adequate to cover the range of the sand sizes used for the experimental work. The sand sample was passed through the series of standard sieves. The weight of sand retained in each sieve was determined, and the cumulative percentage by weight passing through each sieve was calculated. The particle size distribution is shown in Fig. 3.10.

Sand with particle sizes ranging from 0.2 to 0.6 mm were defined as medium sand whereas particle sizes of 0.6−2 mm were considered as

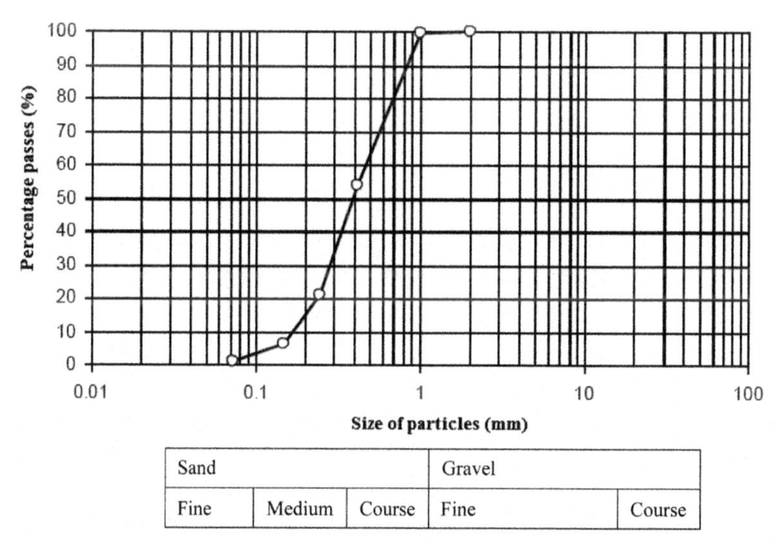

Figure 3.10 Particle size distributions for the sand sample.

Table 3.2 Particle size properties of sand samples
Particle size properties of sand analysis

	Particle size (mm)
D_{10}	0.17
D_{30}	0.32
D_{60}	0.55
C_u	2.8
C_c	1
Percentage of middle sand	45.7%
Percentage of fine sand	54.3%
Percentage of coarse sand	0%

coarse sand. The soils used were therefore classified as uniform medium sand with $D_{50} = 0.50$ mm. The sand properties from particle size distribution analysis are summarized in Table 3.2.

3.7.2 Maximum and Minimum Sand Unit Weight

The minimum dry unit weight represents the loosest condition of a soil mass. The procedure essentially minimized particle segregation. In general, this procedure determined the dry unit weight of oven-dried soil.

The maximum dry unit weight of soil was determined by placing the oven-dried soil mass in a mold and applying a dead weight (1 psi) as a surcharge to the surface of the soil. The mold, soil mass, and dead weight were then vibrated vertically using an electromagnetic vibrating table for a specific time period. The maximum unit weight was obtained by dividing the oven-dried soil mass by its volume.

The relative dry unit weight represents the degree of compactness of a cohesionless soil mass with respect to the loosest and densest conditions as defined in the laboratory steps. It is, therefore, considered that the relative dry unit weight is a good indicator of the state of compactness of a given soil mass.

The maximum and minimum unit weights for the sand samples were obtained through the test designated D4254–91 as recommended in ASTM standards on soil compaction (1993). Soil test in ASTM standards on soil compaction (1993) with designation D4254–91 recommends the standard test method for obtaining the minimum index density/unit weight and calculation of relative density. Three alternative

Table 3.3 Results of standard test methods for unit weights of sand using a vibratory table

Uniform sand	**First sample**	**Second sample**	**Third sample**	**Average unit weight**
Maximum unit weight, γ_{max} (kN/m^3)	17.05	16.9	17.11	17
Minimum unit weight, γ_{min} (kN/m^3)	15.04	14.91	15.02	15

procedures were suggested to determine the minimum unit weight, as follows:

1. Test method A—using a funnel pouring device or a hand scoop to deposit material in the mold.
2. Test method B—depositing material into a mold by extracting a soil-filled tube.
3. Test method C—depositing material by inverting a graduated cylinder.

Test method A is the preferred procedure to be used in conjunction with ASTM standards on soil compaction (1993) designation D4253, whereas test methods B and C are for testing used in conjunction with special studies.

ASTM standards on soil compaction (1993) test designation D4253 are the recommended standard test methods for obtaining maximum unit weight using a vibratory table. Four alternative procedures are suggested to determine the maximum index density and unit weight:

1. Test method 1A—using oven-dried soil and an electromagnetic vertically vibrating table.
2. Test method 1B—using wet soil and an electromagnetic vertically vibrating table.
3. Test method 2A—using oven-dried soil and an eccentric or cam-driven vertically vibrating table.
4. Test method 2B—using wet soil and an eccentric or cam-driven vertically vibrating table.

For the purpose of this research, test method 1A was used based on the available equipment. Results of maximum and minimum unit weights are presented in Table 3.3 for sand samples.

A tolerance layer must be performed in order to determine the unit weight of cohesionless soil under compaction. Generally, the quantity of a known mass of cohesionless soil was poured into the test box until the thickness of the sand layer reached 50 mm, then the sand layer was compacted using a vibrator until the desired settlement was achieved. The settlement can be calculated using the equation below:

$$s = t - \frac{Mg}{A\gamma} \quad (3.1)$$

where
t = initial thickness of cohesionless soil layer (mm)
s = layer settlement (mm)
M = mass of sample (kg)
A = test box area (m^2)
γ = desired unit weight of cohesionless soil (kN/m^3)
g = acceleration of gravity force (9.81 m^2/s)

The desired unit weight of soil was 17 kN/m^3 and the area of the test box was 0.5 m^2. Eq. (3.1) is changed to the following formula:

$$s = 50 - 1.15\,M \quad (3.2)$$

According to above equation, the average time of compaction was 2 minutes for each layer. Based on the time of compaction, each 50-mm-thick layer of cohesionless soil required approximately 2 minutes to achieve 17 kN/m^3.

3.7.3 Direct Shear Test Using Small Shear Box

The direct shear test includes the testing of a square prism of soil that is laterally restrained and sheared along a mechanically involved horizontal plane while being subjected to pressure applied along a plane normal to the shearing plane. The shearing resistance was measured at regular intervals using dataloggers and tests were carried out with four samples using different normal pressures until the shearing resistance reached the maximum sustainable value. The relationship between measured shear stress failure and normal applied stress obtained would enable the effective shear strength parameters and internal friction to be derived.

The normal stress acting on uniform sand in the direct shear test conducted for this research is given in Table 3.4.

Results of the direct shear tests conducted on uniform sand are given in Figs. 3.11 and 3.12. Derivation of the internal friction is obtained from the slope variation of the shear stress τ with the normal stress σ.

A summary of the sand properties and results for the direct shear tests are presented in Table 3.5.

Table 3.4 Normal stress acting on sand sample during direct shear test

	First sample	Second sample	Third sample	Fourth sample
Normal stress (kN/m^2)	27	54	82	112

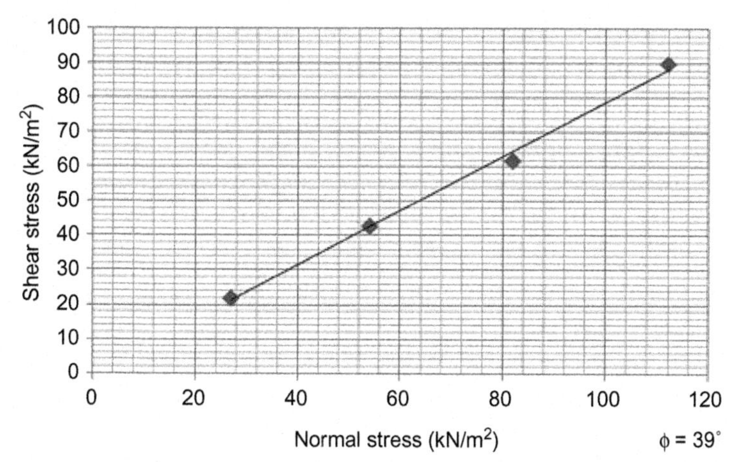

Figure 3.11 Variation of shear stress, τ, versus normal stress, σ, for direct shear tests in loose sand.

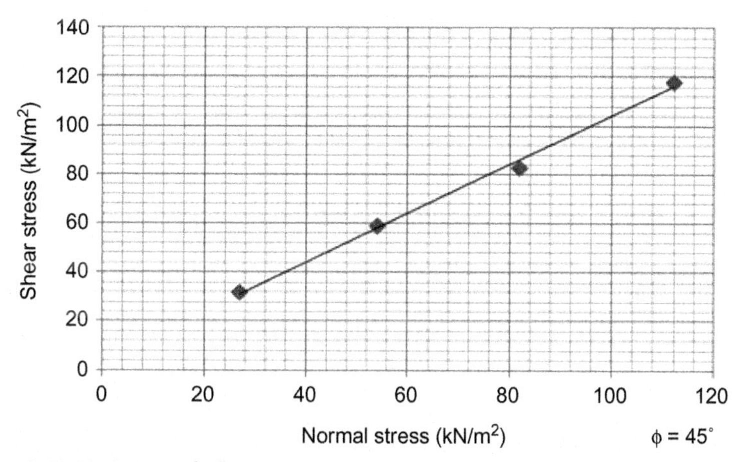

Figure 3.12 Variation of shear stress, τ, versus normal stress, σ, for triaxial tests in dense sand.

Table 3.5 Summary of soil properties and results of Ø from direct shear tests (BS 1377, Part7:1990)

	Loose sand	Dense sand
Unit weight (kN/m^3)	15	17
Ø (degree) in plane strain condition	39	45
Ø (degree) in three-dimensional condition	38	44

3.7.4 Particle Density Using the Small Pyknomoeter Method

The term "particle density" is used instead of the term "specific gravity" in this updated version of this particular standard (BS 1377, Part 2:1990). There are basically three procedures subscribed, namely:

1. The gas jar method: suitable for most soils involving those containing gravel-sized particles.
2. The small pyknometer method: the more definitive method for soils consisting of clay, silt, and sand-sized particles.
3. The pyknometer method: suitable for most soils up to medium gravel size and the least accurate compared with the previous two methods.

The test for particle density was conducted using the small pyknometer method for the purpose of this research as the facilities were readily available and the method was suitable. Tests were performed on three samples for sand and the average value for particle density was used. The results of the test are given in Table 3.6.

3.7.5 Summary of Soil Properties

A summary of soil properties for all tests conducted as described previously is given in Table 3.7.

3.8 ANCHOR PLATE SPECIFICATIONS

The experimental and numerical steps include two categories, namely tests and models of uplift capacity for symmetrical anchor plates of variable shapes and tests for symmetrical anchor plates of variable sizes as

Table 3.6 Summary of particle density test on uniform sand

Bottle no.	First sample	Second sample	Third sample
Mass of bottle (w_1) g	27.30	30.45	29.70
Mass of bottle + sand (w_2) g	34.75	39.10	38.64
Mass of bottle + sand + water (w_3) g	82.35	86.45	85.30
Mass of bottle + water (w_4) g	77.64	81.50	79.80
Mass of sand ($w_2 - w_1$) g	7.49	8.63	8.96
Mass of water whose volume is equivalent to sand $(w_4 - w_1) - (w_3 - w_2)$ g	2.78	3.70	3.46
Particle density of sand $\rho_s = \frac{w_2 - w_1}{(w_4 - w_1) - (w_3 - w_2)}$	2.64	2.63	2.64
Average particle density, ρ_s	2.64		

Table 3.7 Summary of soil properties for tests previously undertaken

Conditions	Uniform sand
Loose	$\varnothing = 38$ degrees $G_s = 2.64$ $\gamma_{min} = 15\ kN/m^3$
Dense	$\varnothing = 44$ degrees $G_s = 2.64$ $\gamma_{max} = 17\ kN/m^3$

Table 3.8 Model properties included in uplift tests for the testing of symmetrical anchor plate size as the parameter under research

Model no.	Material	Anchor plate shape	Dimension (mm)	Embedment ratio (*L/D*)
1	Steel	Square	50	1–4
2	Steel	Square	75	1–4
3	Steel	Square	100	1–4
4	Steel	Circular	50	1–4
5	Steel	Circular	75	1–4
6	Steel	Circular	100	1–4
7	Steel	Rectangular	200 mm × 50 mm	1–4
8	Steel	Rectangular	300 mm × 75 mm	1–4

illustrated in Table 3.8. Uplift tests include symmetrical anchor plates of variable sizes and shapes; Fig. 3.13 illustrates the different symmetrical anchor plates used.

3.9 FAILURE MECHANISM

The failure mechanism tests were performed as shown in Fig. 3.14. In these tests, failure patterns were based on extreme uplift loads and embedment ratios. The aim of these tests was to understand the behavior of loosely and densely packed soil around symmetrical anchor plates due to uplift load.

Failure mechanism tests were carried out in dry loose sand and dense sand with unit weights of 15 kN/m^3 and 17 kN/m^3, respectively. The thickness of each layer was induced with 4 mm dyed that was placed on the front face of test box. Each sand layer was 50 mm. Loading was applied to the anchor plates through the loading cable to all test series in

Figure 3.13 Symmetrical anchor plates used for testing of shapes and sizes parameters subjected to uplift.

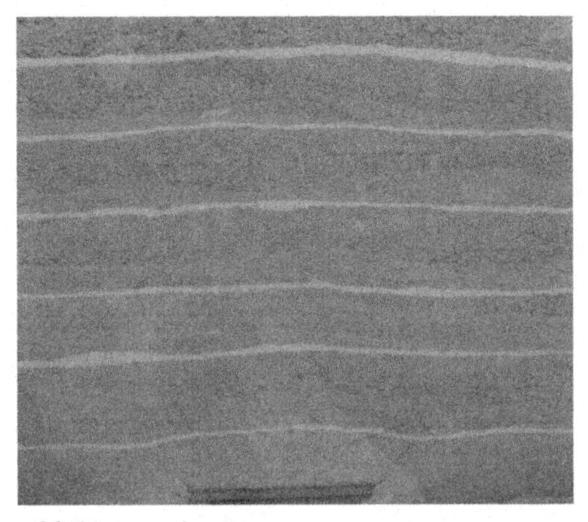

Figure 3.14 Setup of failure mechanism.

sand. The symmetrical anchor plates were made to move until sufficient distance was reached to achieve a clear failure pattern. The results of these tests are given in Chapter 4, Experimental and Numerical Results.

3.10 DETERMINATION OF TEST PARAMETERS

In establishing the physical quantities for breakout factor and uplift load of symmetrical anchor plates in sand, the author assumed a model test without involving the gravity factor. Table 3.9 shows the physical and

Table 3.9 Physical quantities assumed to be included in establishing the uplift capacity and breakout factors

Material properties	Symbol	Units	Fundamental quantities
Symmetrical anchor plates			
Uplift force	*P*	kN	MLT^{-2}
Embedment depth	*L*	m	L
Plate size	*D*	m	L
Sand			
Unit weight	γ	kN/m^3	$ML^{-2}T^{-2}$
Friction angle	Ø	Degree	–
Geogrid			
Depth of first layer of geogrid reinforcement top plate base	*x*	m	L
Vertical distance between geogrid reinforcement layers	*u*	m	L
Geogrid reinforcement length	*B*	m	L
Number of geogrid layers	*N*	–	–
GFR			
Depth of first layer of GFR reinforcement top plate base	*x*	m	L
Vertical distance between GFR reinforcement layers	*u*	m	L
GFR reinforcement length	*B′*	m	L
Number of GFR layers	*N*	–	–
GFR length	*Lg*	m	L
Angle of GFR with horizontal	θ	–	–

fundamental quantities included in the relationship. The same factors were also assumed to be applied to the symmetrical anchor plates. By simulations of appropriate physical quantities, different sets of dimensionless products were developed.

3.10.1 Determination of Test Parameters in Non-reinforced Sand

The main parameters of collapse load which may act on soil parameters are those due to unit weight of sand, internal friction, symmetrical anchor plate's embedded depth, and size of symmetrical anchor plates.

In full-scale model analysis, an equation of these parameters may be expressed in dimensionless quantities as stated as below:

$$f_1(P, L, D, \varnothing, \gamma) = 0 \tag{3.3}$$

f_1 may be expressed as f_2, where,

$$f_2(\pi_1, \pi_2, \pi_3) = 0 \tag{3.4}$$

Since the $\varnothing$ is a dimensionless unit,

$$\pi_1 = \varnothing \tag{3.5}$$

Then,

$$P = f(L, D, \gamma)$$

$$P = L^{\alpha} D^{\beta} \gamma^{c}$$

$$MLT^{-2} = (L)^{\alpha}(L)^{\beta}(ML^{-2}T^{-2})^{c}$$

$$\alpha = 1, \quad \beta = 2, \quad c = 1$$

Then,

$$P = LD^2\gamma$$

$$\pi_2 = P/LD^2\gamma \tag{3.6}$$

L and D have the same dimensional form, thus

$$\pi_3 = L/D \tag{3.7}$$

Thus,

$$f_1\left(\varnothing, P/LD^2\gamma, L/D\right) = 0 \tag{3.8}$$

$$\frac{P}{LD^2\gamma} = f\left(\varnothing, \frac{L}{D}\right)$$

$$P = f\left(\varnothing, \frac{L}{D}\right) \times LD^2\gamma \tag{3.9}$$

Thus the expression in Eq. (3.8) indicates the breakout factor which is a function of the friction angle and the embedment ratio, where P is the ultimate uplift load obtained from test, D is the width of anchor plate, L is the embedded depth of the anchor plate, γ is the dry unit weight, $\varnothing$ is the internal friction angle, and L/D is the embedment ratio. The internal friction angle is constant for the test in non-reinforced sand. These relationships are determined from the experimental results which will be comprehensively discussed in the following chapter.

3.10.2 Determination of Test Parameters in Reinforced Sand Using Geogrid

The main parameters of uplift load which may act on soil parameters and reinforcement materials such as geogrid layers are those due to unit weight of sand, internal friction, symmetrical anchor plate's embedded depth and size of symmetrical anchor plates, geogrid width, number of reinforcement layers, depth of first layer of reinforcement top anchor plate, and vertical distance between geogrid layers. In full-scale model analysis, the equation of those parameters may be expressed in dimensionless quantities as stated as below:

$$f_1(P, L, D, \varnothing, \gamma, x, u, N, B) = 0 \tag{3.10}$$

f_1 may be expressed as f_2,where,

$$f_2(\pi_1, \pi_2, \pi_3, \pi_4, \pi_5, \pi_6) = 0 \tag{3.11}$$

Since the $\varnothing$ is dimensionless unit, thus

$$\pi_1 = \varnothing \tag{3.12}$$

Then,

$$P = f(L, D, \gamma)$$

$$P = L^{\alpha} D^{\beta} \gamma^{c}$$

$$MLT^{-2} = (L)^{\alpha}(L)^{\beta}(ML^{-2}T^{-2})^{c}$$

$$\alpha = 1, \quad \beta = 2, \quad c = 1$$

Then,

$$P = LD^2\gamma$$

$$\pi_2 = P/LD^2\gamma \tag{3.13}$$

x and D have the same dimensional form, so

$$\pi_3 = x/D \tag{3.14}$$

Then, since the N and B are dimensionless units, thus

$$\pi_4 = N \tag{3.15}$$

$$\pi_5 = B \tag{3.16}$$

u and D have the same dimensional form, so

$$\pi_6 = u/D \tag{3.17}$$

Thus,

$$f_1\left(\varnothing, \frac{P}{LD^2\gamma}, \frac{L}{D}, \frac{x}{B}, B, \frac{u}{D}, N\right) = 0 \tag{3.18}$$

Then,

$$\frac{P}{LD^2\gamma} = f\left(\varnothing, \frac{L}{D}, \frac{x}{B}, B, \frac{u}{D}, N\right)$$

$$P = f\left(\varnothing, \frac{L}{D}, \frac{x}{D}, B, \frac{u}{D}, N\right) \times LD^2\gamma \tag{3.19}$$

Thus, the expression in Eq. (3.19) indicates the breakout factor, which is a function of internal friction angle, embedment ratio, geogrid width, number of reinforcement layers, depth of the first layer of reinforcement on top of the anchor plate, number of geogrid layers, and vertical distance between geogrid layers, where P is the ultimate uplift load obtained from test, D is the width of the anchor plate, L is the embedded depth of the anchor plate, γ is the dry unit weight, $\varnothing$ is the internal friction angle, B is the geogrid width, N is the number of geogrid layers, u/D is the ratio of the vertical distance between the geogrid layers and the plate width, x/D is the ratio of approximately the first geogrid layer to the anchor plate, and L/D is the embedment ratio. The internal friction angle is a constraint for the test in non-reinforced sand. These relationships are determined from the experimental results which will be comprehensively discussed in the following chapter.

3.10.3 Determination of Test Parameters in Reinforced Sand Using GFR

The main parameters of uplift load which may act on soil parameters and reinforcement materials such as GFR layers are those due to unit weight of sand, internal friction, symmetrical anchor plate's embedded depth and size of symmetrical anchor plates, geogrid width, number of reinforcement layers, depth of first layer of reinforcement top anchor plate, and vertical distance between geogrid layers. In full-scale model analysis, equations of these parameters may be expressed in dimensionless quantities as given below using Eq. 3.9:

$$f_1(P, L, D, \varnothing, \gamma, x, u, N, B', Lg, e) = 0 \tag{3.20}$$

f_1 may be expressed as f_2, where,

$$f_2(\pi_1, \pi_2, \pi_3, \pi_4, \pi_5, \pi_6) = 0 \tag{3.21}$$

Since the GFR length and GFR angle with reinforcement layers are constant, thus they are constant although $\varnothing$ is a dimensionless unit, thus

$$\pi_1 = \varnothing \tag{3.22}$$

Then,

$$P = f(L, D, \gamma)$$

$$P = L^{\alpha} D^{\beta} \gamma^{c}$$

$$MLT^{-2} = (L)^{\alpha}(L)^{\beta}(ML^{-2}T^{-2})^{c}$$

$$\alpha = 1, \quad \beta = 2, \quad c = 1$$

Then,

$$P = LD^2\gamma$$

$$\pi_2 = P/LD^2\gamma \tag{3.23}$$

x and D have the same dimensional form, thus

$$\pi_3 = x/D \tag{3.24}$$

Then, since the N and B are dimensionless units,

$$\pi_4 = N \tag{3.25}$$

$$\pi_5 = B' \tag{3.26}$$

u and D have the same dimensional forms, so

$$\pi_6 = \frac{u}{D} \tag{3.27}$$

Thus,

$$f_1\left(\varnothing, \frac{P}{LD^2\gamma}, \frac{L}{D}, \frac{x}{B}, B, \frac{u}{D}, N\right) = 0 \tag{3.28}$$

Then,

$$\frac{P}{LD^2\gamma} = f_1\left(\varnothing, \frac{L}{D}, \frac{x}{B}, B, \frac{u}{D}, N\right)$$

$$P = f\left(\varnothing, \frac{L}{D}, \frac{x}{B}, B', \frac{u}{D}, N\right) \times LD^2\gamma \tag{3.29}$$

Thus the expression in Eq. (3.29) indicates the breakout factor, which is a function of internal friction angle, embedment ratio, GFR width, number of reinforcement layers, depth of the first layer of reinforcement on top of the anchor plate, number of GFR layers, and vertical distance between GFR layers, where P is ultimate uplift load obtained from the test, D is width of anchor plate, L is the embedded depth of anchor plate, γ is the dry unit weight, $\varnothing$ is the internal friction angle, B' is the GFR width, N is the number of GFR layers, u/D is the ratio of the vertical distance between the GFR layers and the plate width, x/D is the ratio of approximately the first GFR layer to the anchor plate, and L/D is the embedment ratio. The internal friction angle is the constraint for the test in non-reinforced sand. These relationships are determined from the experimental results which will be comprehensively discussed in the following chapter.

3.11 FINITE ELEMENT METHOD

A series of two-dimensional finite element analyses (FEA) on a prototype symmetrical-anchor-plate–sand system was performed in order to assess the experimental model tests results and to ascertain the deformation behavior within the sand body. The analysis was performed using the finite element program, PLAXIS (professional version 8, Bringkgreve and Vermeer, 1998). PLAXIS is a geotechnical software package that can analyze the soil problems. In general, the initial conditions comprise the initial groundwater conditions, the initial geometry configuration, and the initial effective stress state. The sand layer in this research was dry, so there was no need to enter groundwater conditions. The analysis was performed by means of the Hardening Soil Model (HSM). The geometry of the prototype anchor-plate–box system was supposed to be the same as the experimental model. The same gradient of model test and the material of the steel plate for the symmetrical anchor plate, geogrid, GFR and sand was used in the prototype research. The software enables the automatic production of 6 or 15 node triangle plane strain elements for square and rectangular plates and axisymmetry for circular plates in the sand. Tables 3.10–3.12 tabulate the preparation of the sand, the geogrid, and the plate used in the analysis, respectively.

Table 3.10 Material properties used in PLAXIS

Parameter value	Loose packing	Dense packing
Cohesion, c (kPa)	0.5	0.5
Residual angle of internal friction (degree)	38	44
Angle of dilatancy (Ψ°)	8	14
Unit weight, γ (kN/m^3)	15	17
Secant stiffness, E_{50} (kN/m^2)	20,000	30,000
Initial stiffness, E_{OED} (kN/m^2)	20,000	30,000
Unloading/reloading stiffness, E_{UR} (kN/m^2)	60,000	90,000
Poisson's ratio	0.2	0.2
Power for stiffness stress dependency (m)	0.5	0.5
At rest earth pressure coefficient, K_0	0.38	0.32
R_{inter}	0.9	0.9

Table 3.11 Geosynthetic properties

Geosynthetic type	Geogrid
Polymer type	Polypropylene
Geogrid shape	Square
Apparent opening size	21 mm × 21 mm
Tensile strength (kN/m)	60

Table 3.12 Steel plate properties

Type	**Steel plates**
EI	163 kNm^2/m
EA	3.4×10^5 kN/m

EA and EI are the axial and bending stiffnesses.

A variety of sand models were made in the computer code chosen for this research. The HSM was chosen to model the sand for its analysis, practical importance, and the availability of the parameters needed. The interaction between the symmetrical anchor plates, geogrid, and sand was modeled by means of interface elements, which enabled the specification of a decreased wall friction compared to the soil friction. The parameters used for numerical simulation are depicted in Tables 3.10 and3.11. The model geometry based on the finite element method by means of PLAXIS verified for the analysis is shown in Fig. 3.15. The left vertical line of the geometry model was constrained horizontally but the bottom horizontal boundary was constrained in both the horizontal and vertical directions. The sand parameters determined for the top and bottom soil layers were assumed to be similar to all parts of unreinforced sand. For the reinforced sand, a reinforcement layer was placed at the mentioned depth although the suitable strength for reduction factors between the contact surfaces and stiffness of the geosynthetics is considered. The prescribed load was loaded in increments accompanied by iterative analysis up to failure. The boundary conditions showed that the vertical boundary is free vertically and constrained horizontally until the bottom horizontal boundary is completely fixed. The program can be the automatic product of six node triangle

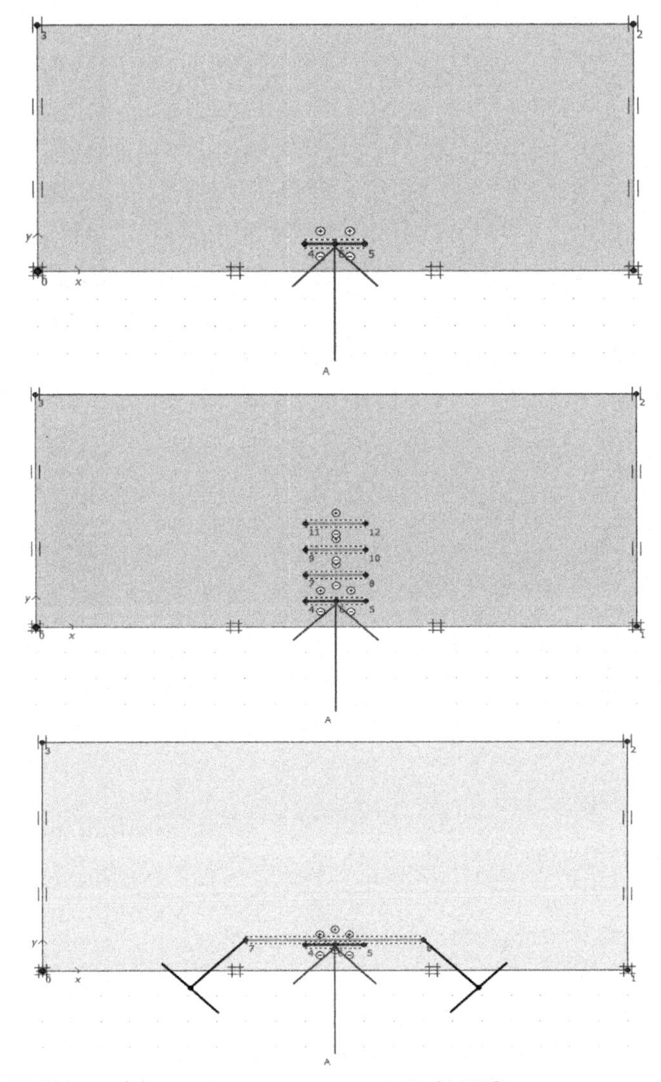

Figure 3.15 The model geometry on a prototype in PLAXIS.

plane strain elements for the sand and three node tensile elements for the symmetrical anchor plate, geogrid, and GFR. The analyzed geometry and the produced mesh and related boundary conditions are shown in Fig. 3.16. The symmetrical anchor plates were modeled with plate item in PLAXIS. The GFRs were simulated by fixed-end anchors.

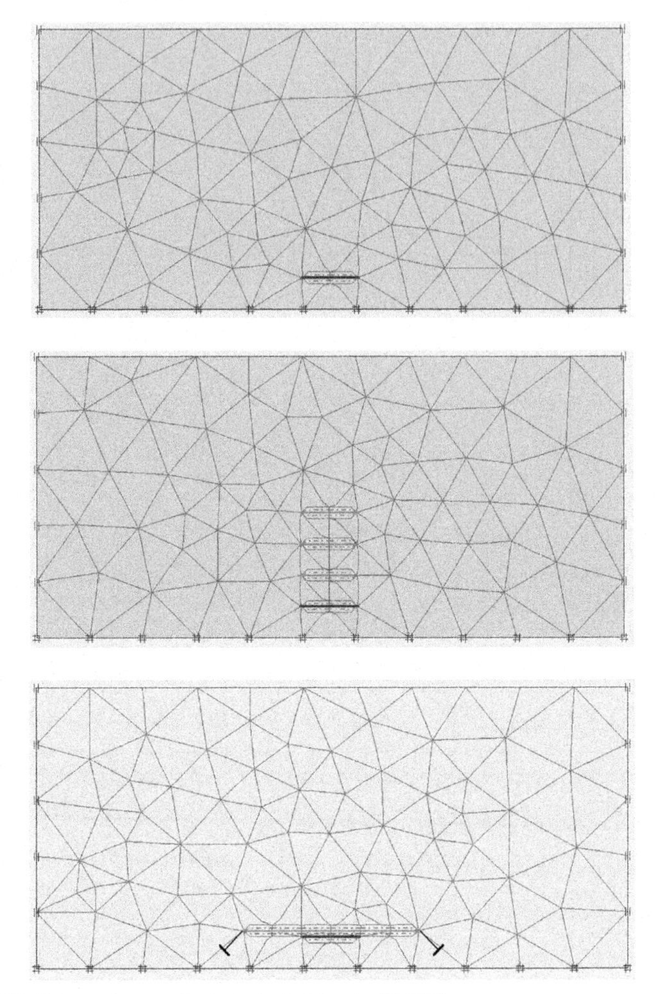

Figure 3.16 Geometry mesh on a prototype in PLAXIS.

The fixed-end anchors were springs which were used to simulate the tying of a single point. Fixed-end anchors are shown as rotated T-shapes, based on specific angles.

REFERENCE

Bringkgreve R, Vermeer P (1998) PLAXIS-finite element code for soil and rock analysis, version 7. Plaxis BV, The Netherlands.

CHAPTER 4

Experimental and Numerical Results

4.1 INTRODUCTION

This chapter presents the results of the uplift responses based on experimental test series and numerical analysis, as discussed in the previous chapters. The numerical analyses were performed using PLAXIS as a finite element program. The research involved the numerical and experimental aspects of the uplift capacity of symmetrical anchor plates during uplift tests and their displacement on non-reinforced, reinforced, and grid fixed reinforcement (GFR) reinforced soil. Each aspect of the research is divided into the test series to enable a more systematic approach towards the analysis. Sand was used as an embedment medium in this research. In order to obtain a criterion for density, loosely and densely packed sand were used. Loose packing was achieved by sand raining methods, while dense packing was achieved through hand compaction via electrical vibrator. Results obtained from tests conducted using numerical and experimental results will be discussed in this chapter. The effect of embedment ratio, sand density, shape factor, breakout factor, and failure mechanism patterns of models are detailed in non-reinforced, reinforced, and GFR-reinforced soil in numerical and experimental tests. The failure mechanism patterns of models in loose sand and dense sand observed based on experimental and numerical analysis are presented in this chapter.

4.2 UPLIFT TEST: OVERVIEW

Tables 4.1–4.3 present a summary of the results for uplift tests of symmetrical anchor plates based on non-reinforced, reinforced, and GFR soil through simulation and experimental work. The details of every uplift experiment and numerical simulation are illustrated in Appendix B in terms of the uplift-force–vertical-displacement relationship.

Appendix B describes the relationship between the uplift capacity and the vertical displacement response of symmetrical anchor plates tested in

Soil Reinforcement for Anchor Plates and Uplift Response.
DOI: http://dx.doi.org/10.1016/B978-0-12-809558-4.00004-2
© 2017 Elsevier Inc.
All rights reserved.

Table 4.1 Summary of uplift capacity result for test series 1–4 and 11($L/D = 4$)

Symmetrical anchor plate	**Uplift capacity in non-reinforced sand (*N*)**			
Types	**Loose sand**		**Dense sand**	
	Lab	**PLAXIS**	**Lab**	**PLAXIS**
Circular				
Diameter = 5 cm	150	184	254	220
Diameter = 7.5 cm	339	414	576	498
Diameter = 10 cm	644	788	1,108	956
Square				
Length = 5 cm	152	186	256	222
Length = 7.5 cm	342	420	579	502
Length = 10 cm	640	780	1088	940
Rectangular				
Length = 20 cm	5,576	6,800	9,512	8,200
Length = 30 cm	15,324	18,969	25,728	21,870

Table 4.2 Summary of uplift capacity result for test series 5–7 and 11 ($L/D = 4$)

Symmetrical anchor plate	**Uplift capacity in reinforced sand (*N*)—$B/D = 2$**							
Position	**$x/D = 0$**		**$x/D = 0.5$**		**$x/D = 0.75$**		**$x/D = 1$**	
Types	**Dense**	**Loose**	**Dense**	**Loose**	**Dense**	**Loose**	**Dense**	**Loose**
Circular								
Diameter = 10 cm (Exp)	1284	744	1296	768	1312	748	1324	756
Diameter = 10 cm (Num)	1108	908	1120	940	1132	916	1144	940
Square								
Length = 10 cm (Exp)	1316	768	1292	756	1296	760	1348	776
Length = 10 cm (Num)	1136	940	1116	924	1124	928	1164	948
Rectangular								
Length = 20 cm (Exp)	4560	2712	4368	2568	4410	2568	4428	2538
Length = 20 cm (Num)	3966	3312	3768	3138	3804	3132	3822	3096

Table 4.3 Summary of uplift capacity result for test series 8–11 ($L/D = 4$)

Symmetrical anchor plate	Uplift capacity in grid fixed reinforced sand (N)—$B'/D = 2$							
Position	$x/D = 0$		$x/D = 0.5$		$x/D = 0.75$		$x/D = 1$	
Types	**Dense**	**Loose**	**Dense**	**Loose**	**Dense**	**Loose**	**Dense**	**Loose**
Circular								
Diameter = 10 cm (Exp)	1280	744	1268	740	1292	744	1316	764
Diameter = 10 cm (Num)	1104	912	1096	904	1116	912	1136	936
Square								
Length = 10 cm (Exp)	1316	764	1264	744	1300	764	1344	784
Length = 10 cm (Num)	1136	936	1092	912	1124	932	1160	960
Rectangular								
Length = 20 cm (Exp)	4596	2712	4386	2568	4404	2580	4440	2550
Length = 20 cm (Num)	3966	3312	3786	3132	3798	3150	3828	3114

the experimental and numerical simulation programs. The general trend indicates that soil experiences plastic deformation when symmetrical anchor plates are pulled out. The symmetrical anchor plate uplift force versus the vertical displacement response varied slightly depending on other parameters considered during the tests.

A more detailed discussion and critical examination of the uplift results is described in the following sections. The influence of test parameters that was studied in the 11 test series are discussed concurrent with the peak uplift capacity of symmetrical anchor plates, the symmetrical anchor plate's vertical displacement at uplift test in non-reinforced, reinforced, and GFR-reinforced sand.

4.3 DISCUSSION ON UPLIFT CAPACITY OF SYMMETRICAL ANCHOR PLATES IN NON-REINFORCED SAND

The discussion on uplift capacity deals with the parameters of symmetrical anchor plate's sizes, shapes, sand packing, and embedment ratio separately. This is to enable an impartial and focused review of the effects of each parameter on the symmetrical anchor plate during uplift in non-reinforced sand. Examples of the calculation method utilized for breakout factor is shown in Appendices C–O.

4.3.1 Test Series 1: Influence of Different Sizes of Symmetrical Anchor Plates on Their Uplift Capacity in Non-reinforced Cohesionless Soil

Fig. 4.1 shows that symmetrical anchor plates experienced an increase in their uplift capacity for every increase in their size. The increase was, however, non-linear for loosely and densely packed soil. Fig. 4.2 shows circular and square anchor plates exhibiting non-linear increase in their uplift capacity in loosely packed soil as their size increases. Fig. 4.3 shows the uplift response of a rectangular anchor plate in loose and dense sand. Fig. 4.1 also shows that the experimental result is higher than numerical modeling for dense sand but lower for loose sand.

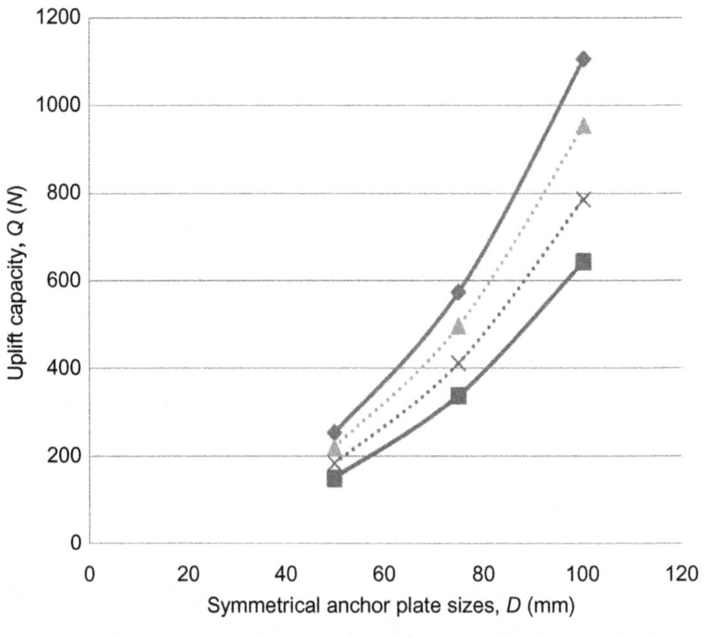

Figure 4.1 Variation in uplift capacity Q with symmetrical anchor plate size D for a circular anchor plate at $L/D = 4$ in loose and dense sand.

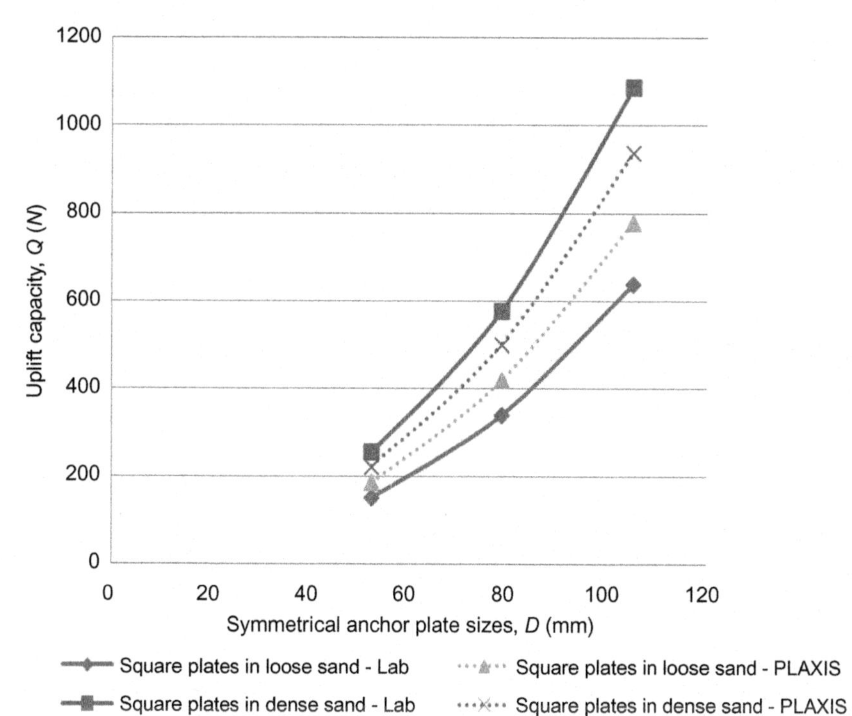

Figure 4.2 Variation in uplift capacity Q with symmetrical anchor plate size D for a square anchor plate at $L/D = 4$ in loose and dense sand.

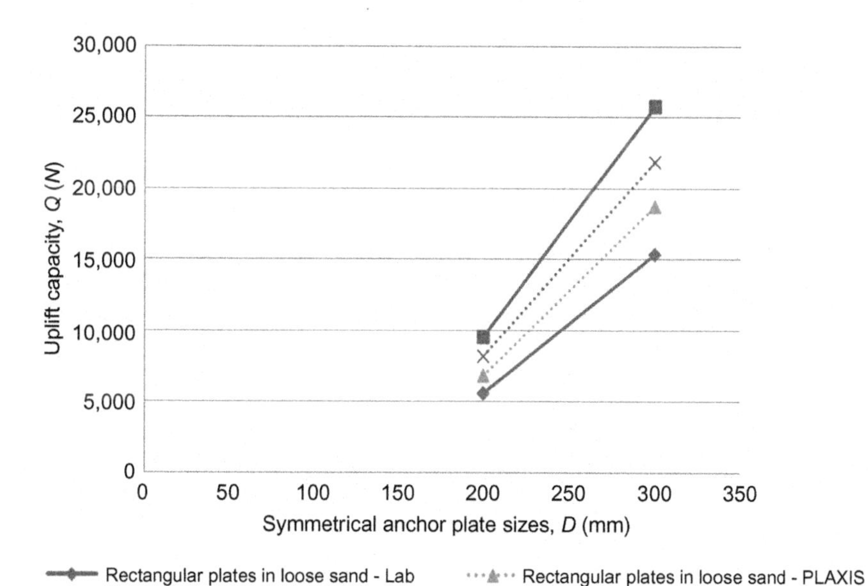

Figure 4.3 Variation in uplift capacity Q with symmetrical anchor plate size D for a rectangular anchor plate at $L/D = 3$ in loose and dense sand.

The increase in the uplift capacity with symmetrical anchor plate size regardless of the packing condition of the soil is due to the increase in lateral stress acting on the symmetrical anchor plates. The lateral stress depends on the depth and anchor plate size. This can be understood from the formula of the uplift response, γAH, where symmetrical anchor size is one of the parameters. As such, an increase in symmetrical anchor plate's size would increase the uplift capacity as given in the formula by Balla (1961) as γAH.

Figs. 4.2 and 4.3 show a geometric increase in the uplift capacity in square and rectangular plates in both loosely and densely packed soil.

4.3.2 Test Series 2: Influence of Symmetrical Anchor Plate Shape on the Uplift Capacity in Non-reinforced Cohesionless Soil

With regard to Figs. 4.4 and 4.5, an increase in symmetrical anchor plate size causes a non-linear increase in the range of uplift response in both loose and dense packing. Referring to Figs. 4.4 and 4.5, comparison of uplift capacities based on symmetrical anchor plate shapes shows that symmetrical circular anchor plates have higher uplift capacities compared to symmetrical square anchor plates for both loose and dense packing. A symmetrical anchor plate subjected to uplift causes extensive shear forces to develop adjacent to the symmetrical anchor plate. This can be understood from the formula of the uplift response γAH, where symmetrical anchor shape is one of the parameters because Balla (1961) used a correlation factor for changing the uplift response from square to circular or rectangular or vice versa. The increasing difference in uplift response between square, circular and rectangular plates with increasing symmetrical anchor plate's sizes are to be expected since uplift capacity increases follow a geometric progression with increasing depth. As such, a deeply embedded symmetrical rectangular anchor plate would be substantially more resistant to uplift forces than a symmetrical square or circular anchor plate due to the geometric progression in capacity with increasing symmetrical anchor plate depth. This behavior is in agreement with findings by Balla (1961), Meyerhof and Adams (1968), and Vesic (1971) and studies on sand—symmetrical-anchor-plate interaction by related researchers.

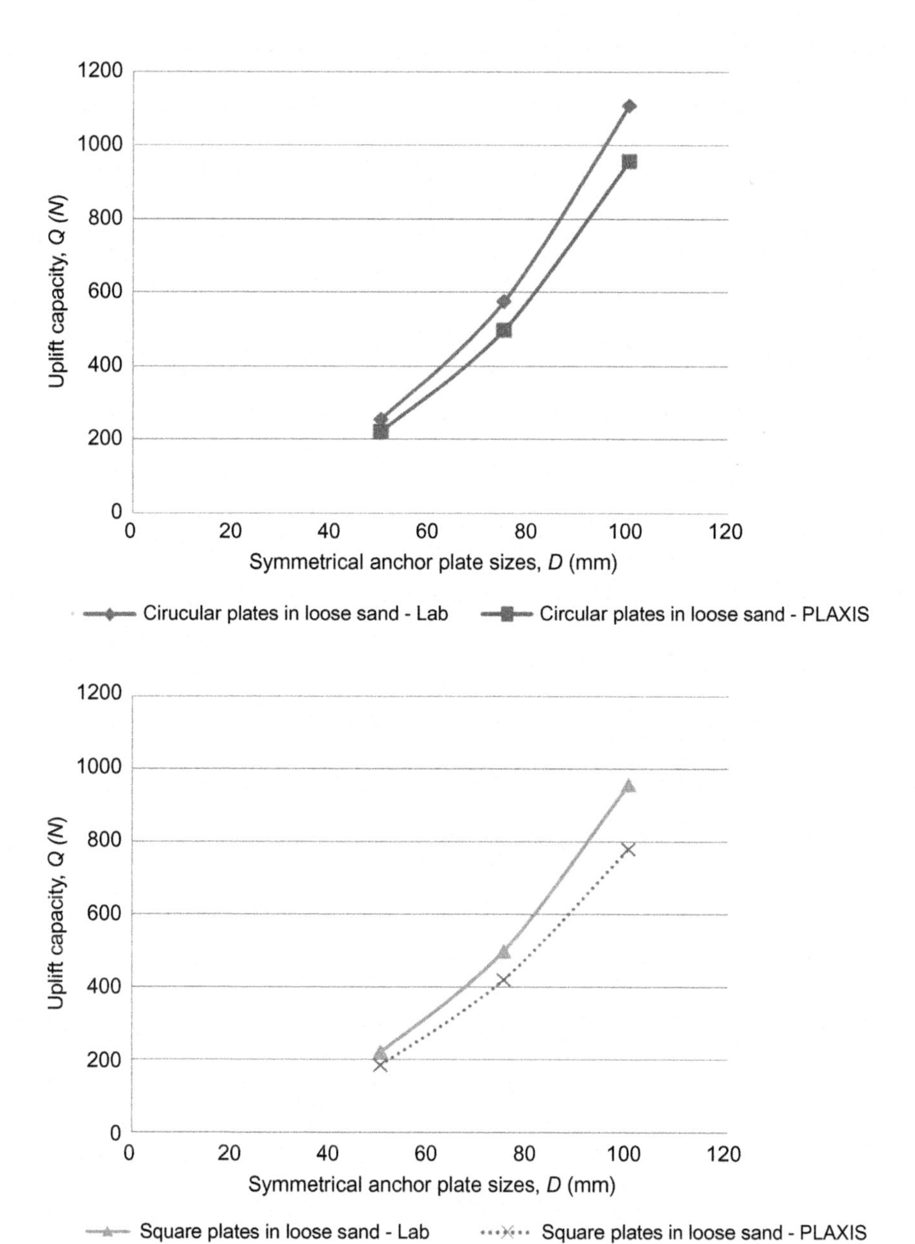

Figure 4.4 Variation in uplift capacity Q with symmetrical anchor plate size D for symmetrical anchor plates at $L/D = 4$ in loose sand.

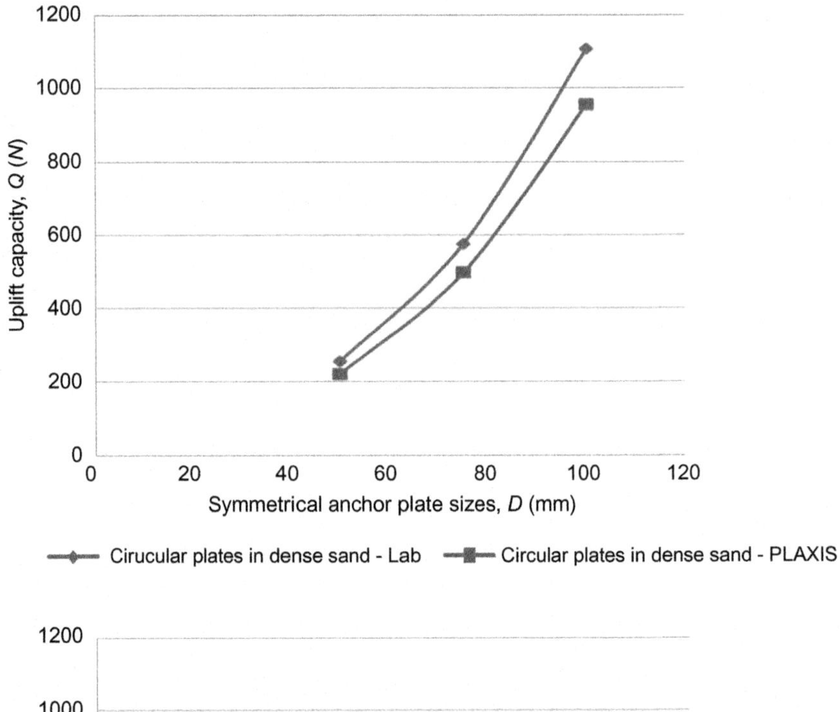

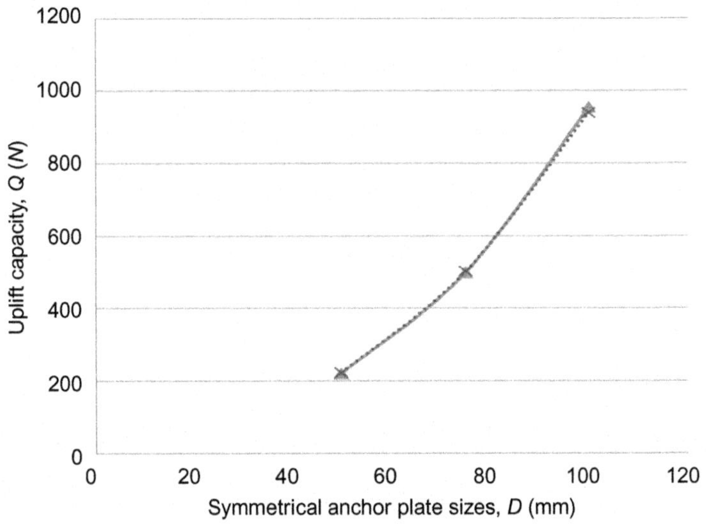

Figure 4.5 Variation in uplift capacity Q with symmetrical anchor plate size D for symmetrical anchor plates at $L/D = 4$ in dense sand.

4.3.3 Test Series 3: Influence of Sand Unit Weight on the Uplift Capacity in Non-reinforced Cohesionless Soil

Unit weights of 15 and 17 kN/m^3 were determined as the value for loose and dense packing for sand, respectively. As the influence of soil packing involves the correlation between test series 1, 2, 3, and 11, discussion of the results are made according to the separate tests. In general, an increase in sand unit weight causes an increase in the uplift capacity of symmetrical anchor plates. From Figs. 4.1 to 4.3, symmetrical anchor plates in dense packing exhibit higher uplift capacity values for sand.

The increases in uplift capacity with sand packing is due to an increase in soil unit weight, γ, which in turn serves to increases lateral stress acting on symmetrical anchor plates and ultimately the uplift capacity γAH. The increases in uplift capacity with the asymmetrical anchor plate's size are related to the parameters included during the calculation of uplift capacity; which in this case is the unit weight.

These findings are similar to those of Balla (1961) and Meyerhof and Adams (1968), who indicated significant differences in uplift capacity values between loose and dense packing.

4.3.4 Test Series 4: Influence of Anchor Plate's Embedded Ratio on the Uplift Capacity in Non-reinforced Cohesionless Soil

Analysis of this test series is correlated with test series 1–4 and 11. As shown in Figs. 4.6–4.12, symmetrical anchor plates at maximum embedment ratio, $L/D = 4$, had higher uplift capacities than symmetrical anchor plates at minimum embedment ratio, $L/D = 1$ in both results of experimental and numerical modeling. The increases were, however, non-linear for loosely and densely packed soil.

4.4 DISCUSSION ON UPLIFT CAPACITY OF SYMMETRICAL ANCHOR PLATES IN REINFORCED SAND

The discussion on the uplift capacity of symmetrical anchor plates in reinforced soil involves a separate analysis of various parameters, such as the number of geogrid layers, the length of the geogrid layers, and vertical spacing of the geogrid layers. Typical physical and technical properties of geogrid obtained from the manufacturer's data sheet are given in Table 3.11. This enables an impartial and focused review of the effect that each parameter has on the symmetrical anchor plates during the uplift

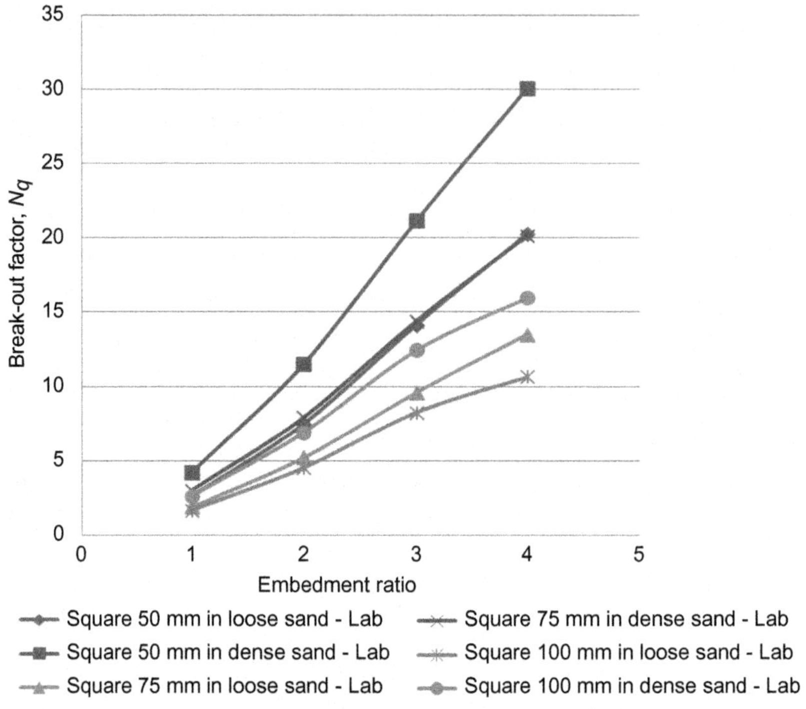

Figure 4.6 Variation of breakout factor N_q with embedment ratio *L/D* for symmetrical square anchor plates in both loose and dense packing in the laboratory.

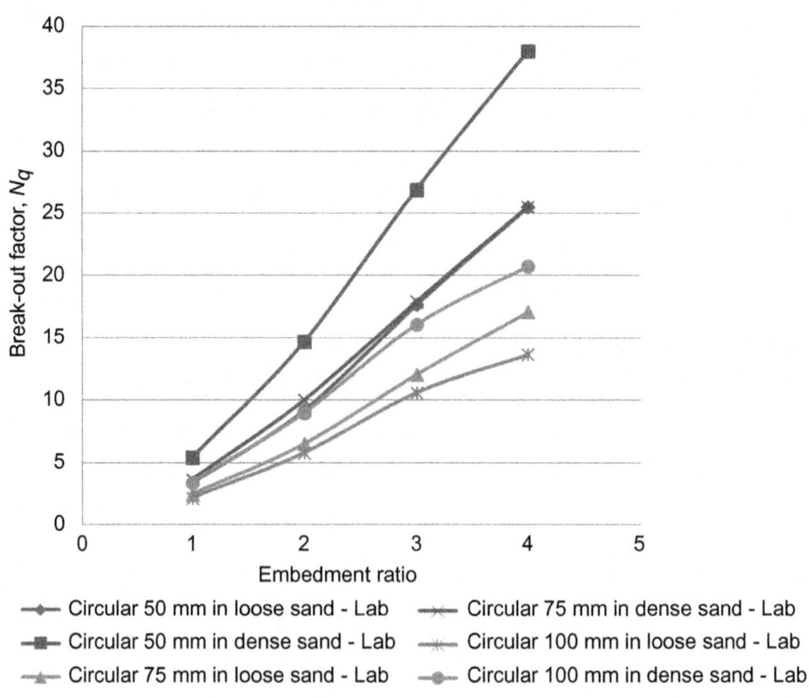

Figure 4.7 Variation of breakout factor N_q with embedment ratio *L/D* for symmetrical circular anchor plates in both loose and dense packing in the laboratory.

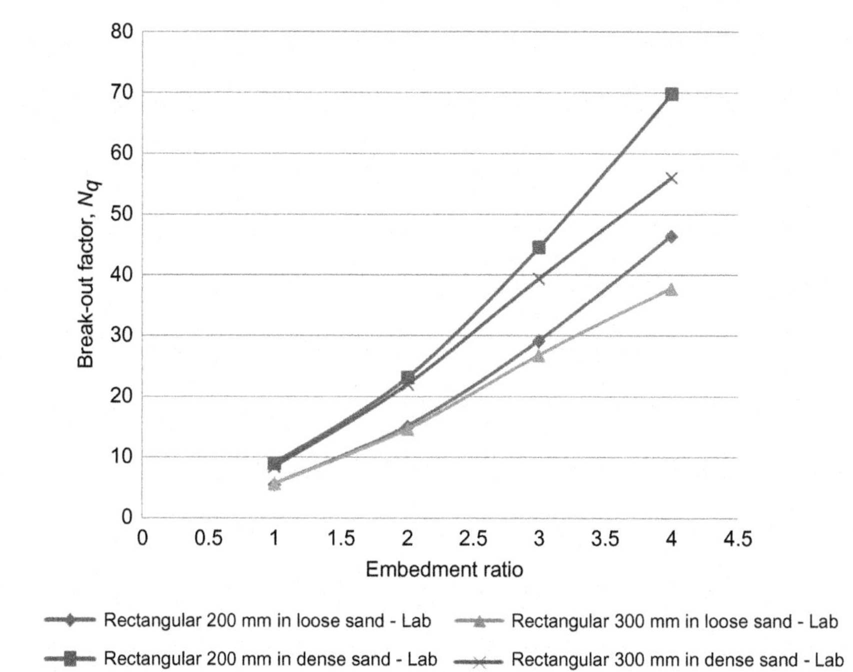

Figure 4.8 Variation of breakout factor N_q with embedment ratio *L/D* for symmetrical rectangular anchor plates in both loose and dense packing in the laboratory.

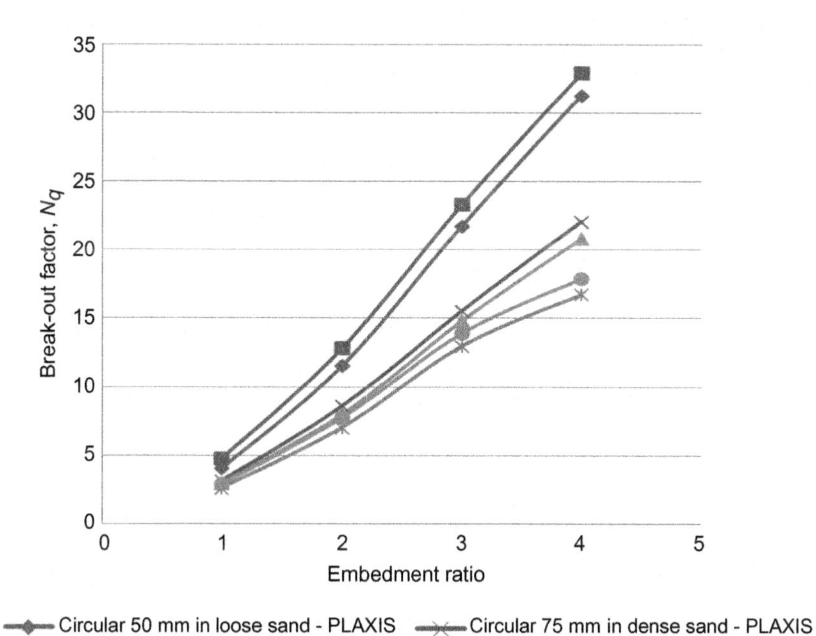

Figure 4.9 Variation of breakout factor N_q with embedment ratio *L/D* for symmetrical circular anchor plates in both loose and dense packing in PLAXIS.

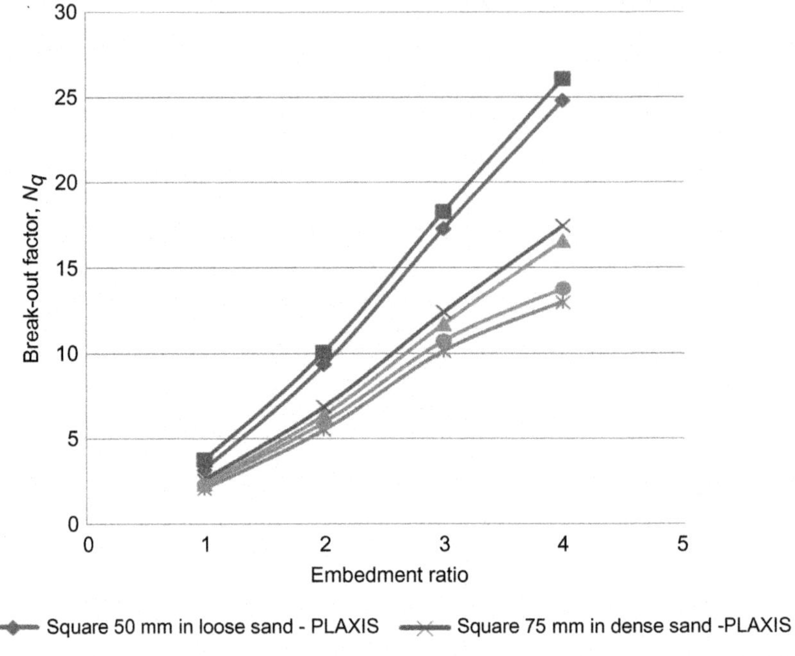

Figure 4.10 Variation of breakout factor N_q with embedment ratio *L/D* for symmetrical square anchor plates in both loose and dense packing in PLAXIS.

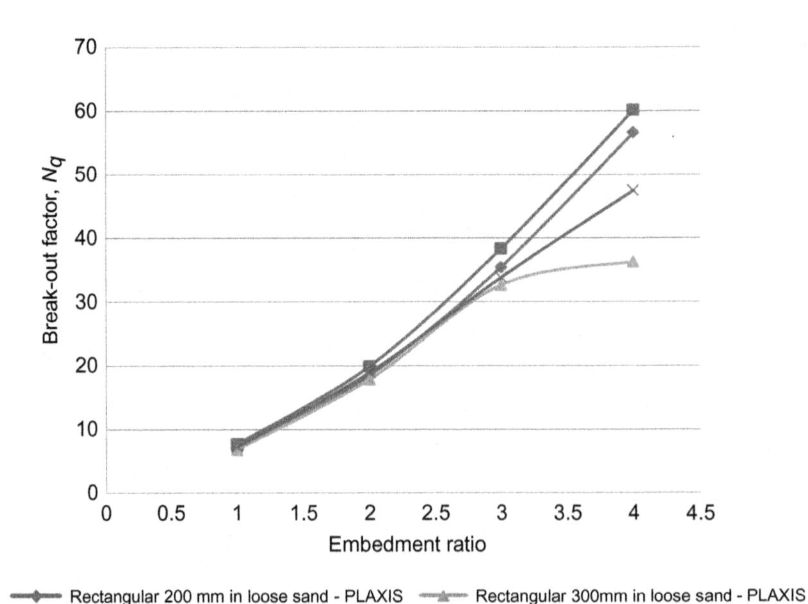

Figure 4.11 Variation of breakout factor N_q with embedment ratio *L/D* for symmetrical rectangular anchor plates in both loose and dense packing in PLAXIS.

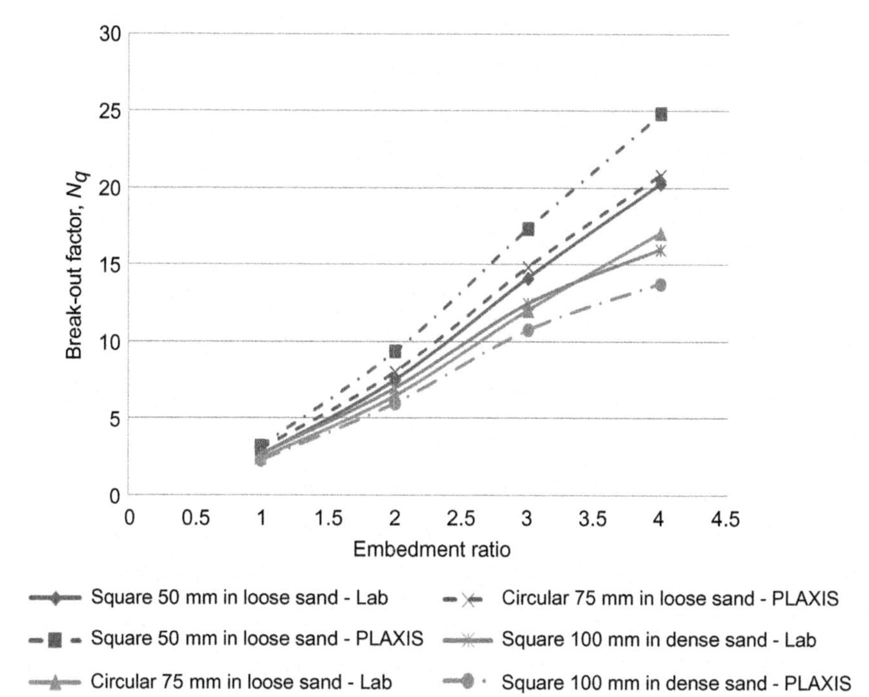

Figure 4.12 Variation of breakout factor N_q with embedment ratio *L/D* for symmetrical anchor plates in both loose and dense packing.

test. One of the areas that needs to be considered is the effect of reinforced soil on symmetrical anchor plates. These components experienced elongation of a certain magnitude, which is calculated in Appendix B.

The uplift response of the symmetrical anchor plates with and without soil reinforcement P_u and P_o were obtained from the uplift-load–displacement curves. The symmetrical anchor plate's capacity improvement due to soil reinforcement is represented using a nondimensional factor, called the symmetrical anchor capacity ratio (ACR) to assist in comparing the test results. ACR is defined as the ratio of the symmetrical anchor plate's ultimate capacity with soil reinforcement, $P_{u\text{-reinforced}}$, to the ultimate capacity of the symmetrical anchor plate in tests without soil reinforcement, P_o. These results are discussed in the following sections. In this research, geogrid length (B') was not kept constant and the number of geogrid layers was varied for research. The variation of ACR with normalized layer spacing, $x/D = 0$ and 0.5 in various B/D are illustrated in Figs. 4.13 and 4.14.

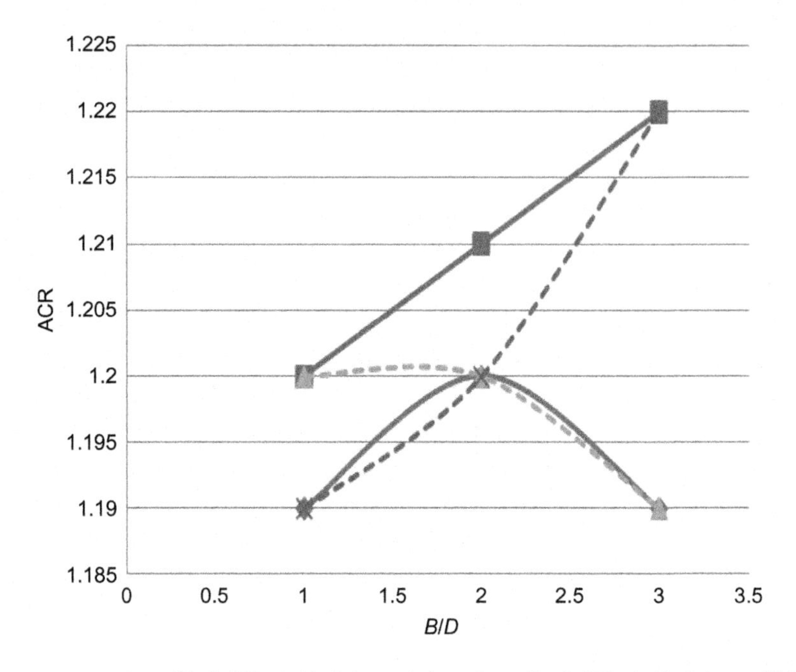

Figure 4.13 Variation of ACR with *B/D* of geogrid layer at $x/D = 0$.

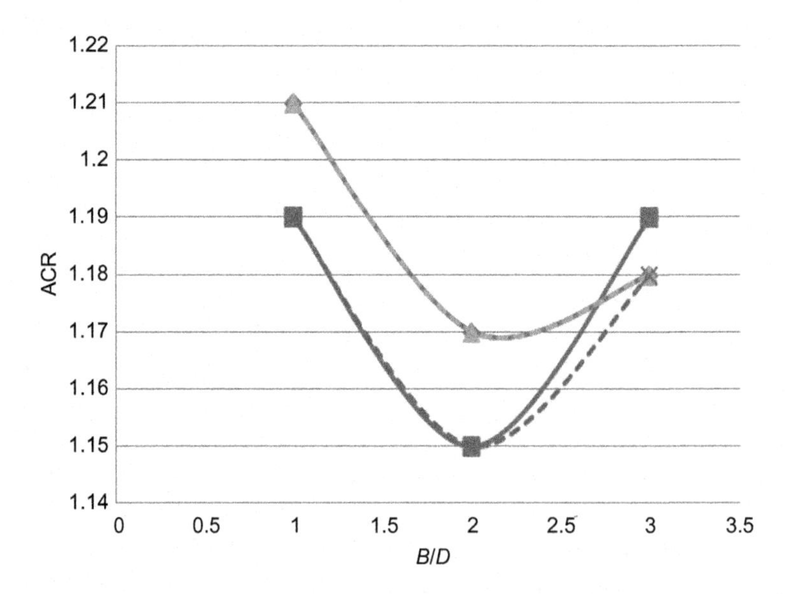

Figure 4.14 Variation of ACR with *B/D* of geogrid layer at $x/D = 0.5$.

4.4.1 Test Series 5: Influence of Number and Vertical Spacing of Geogrid Layers on the Uplift Capacity

With reference to Figs. 4.15 and 4.16, seven tests (1–7 and 11) were performed to research the effects of reinforced sand with various numbers of geogrid inclusions on the behavior of the symmetrical anchor plate located in loose and dense sand. In the reinforced tests, geogrid layers were placed at equal vertical spacings of 0.5*D*, 0.75*D*, and 1*D*, with the first layer resting on the plate between 0*D* and 0.5*D*. The variations in the symmetrical anchor plate's capacity with u/D for a various number of geogrid layers are plotted in Figs. 4.15 and 4.16. The figure clearly shows that the anchor plate behavior much improves with soil reinforcement. Also, it can be seen that inclusion of geogrid layers is much better than no reinforcement. However, the load–displacement ratio curves for a different number of geogrid layers show that the number of geogrid layers has no significant effect on the symmetrical anchor plates' response. In

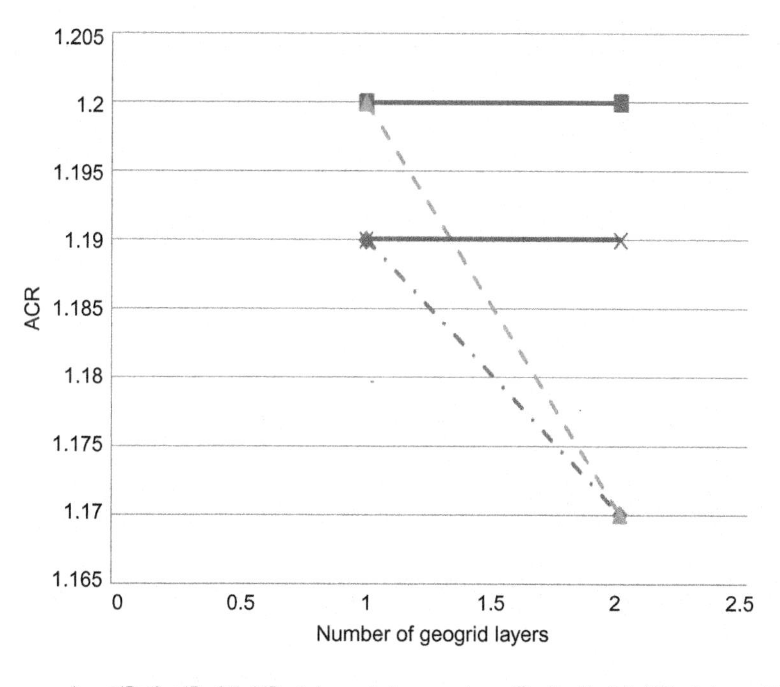

Figure 4.15 Variation of ACR with number of geogrid layers at $x/D = 0$ and $u/D = 0.5$.

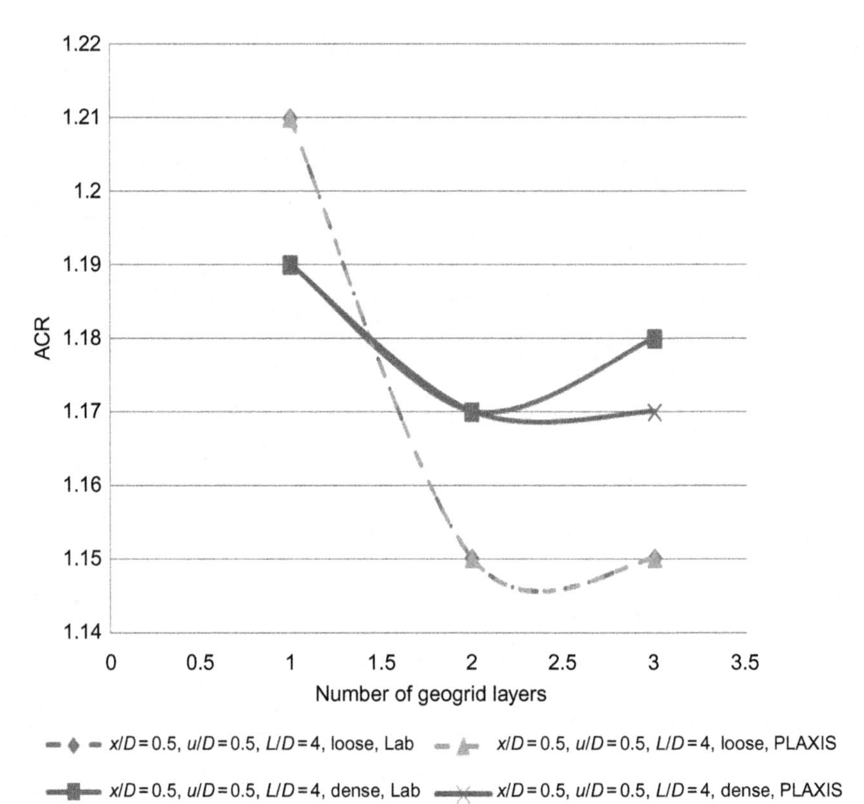

Figure 4.16 Variation of ACR with number of geogrid layers at $x/D = 0.5$ and $u/D = 0.5$.

fact, adding another geogrid layer directly on top of the anchor plate approximately produces the same effect as that produced when multiple geogrid layers are added. Therefore, it was concluded that, in terms of symmetrical anchor plate's capacity, using one geogrid layer is better and more economical than reinforcing the soil itself with several layers. The reason is because both conditions have the same angle of failure zone. Hence, it was decided to carry out the test program on the response of anchor plates adjacent to loose or dense sand using one layer of geogrid placed in the symmetric state over the plate.

4.4.2 Test Series 6: Influence of Geogrid Layer Proximity to the Anchor on the Uplift Capacity

An overview of Tables 4.4 and 4.5 indicates that many series of tests and simulations were performed on symmetrical anchor plates located in loose

Table 4.4 Variation of ACR with influence of geogrid layer proximity to the anchor at $B/D = 1$

Shapes	Embedment ratio	Test type	ACR based on geogrid layer proximity to the symmetrical anchor plate in $B/D = 1$							
	L/D	PLAXIS/laboratory	0		0.5		0.75		1	
Circular (10 cm)	4	Laboratory	1.17	1.16	1.16	1.15	1.16	1.15	1.16	1.15
Circular (10 cm)	4	PLAXIS	1.17	1.15	1.16	1.15	1.16	1.14	1.16	1.15
Square (10 cm)	4	Laboratory	1.20	1.19	1.18	1.18	1.20	1.18	1.20	1.17
Square (10 cm)	4	PLAXIS	1.19	1.20	1.18	1.18	1.20	1.18	1.19	1.18

Table 4.5 Variation of ACR with influence of geogrid layer proximity to the anchor at $B/D = 2$

Shapes	Embedment ratio	Test type	ACR based on geogrid layer proximity to the symmetrical anchor plate in $B/D = 2$							
	L/D	PLAXIS/laboratory	0		0.5		0.75		1	
Circular (10 cm)	4	Laboratory	1.15	1.15	1.16	1.19	1.18	1.16	1.19	1.17
Circular (10 cm)	4	PLAXIS	1.15	1.15	1.17	1.19	1.18	1.16	1.19	1.19
Square (10 cm)	4	Laboratory	1.20	1.20	1.18	1.18	1.19	1.18	1.23	1.21
Square (10 cm)	4	PLAXIS	1.20	1.20	1.18	1.18	1.19	1.18	1.23	1.21

and dense sand with the inclusion of one geogrid layer placed at various distances of 0, $0.5D$, $0.75D$, and $1D$ with $B/D = 1$ over the symmetrical anchor plate. Tables 4.4 and 4.5 show the variation of uplift response with x/D with $B/D = 2$ for the reinforced sand. The improvements in the symmetrical anchor plate's capacity decrease as the geogrid layer is moved away from the plate. Maximum gain in symmetrical anchor plate capacity is attained when the layer is placed resting directly on top of the anchor plate. This may be explained by the fact that the closeness of the geogrid layer to the anchor plate offers greater resistance to anchor displacement, leading to greater failure wedge and larger failure surface area.

4.5 DISCUSSION ON UPLIFT CAPACITY OF SYMMETRICAL ANCHOR PLATES IN GFR-REINFORCED SAND

The uplift capacity of symmetrical anchor plates in GFR-reinforced sand deals with the parameters of number of GFR layers, GFR layer length, GFR layer proximity to the anchor, and vertical spacing of the GFR layers separately. GFR is a new system that ties the geosynthetics into the ground; it is made of a fiber-reinforced polymer strip and an end ball, which tie up to the geosynthetic material and are anchored into the soil as it is layered. Geogrid, as one type of geosynthetic was used in this research, to enable an impartial and focused review of the effects of each parameter on the symmetrical anchor plate at uplift test. One of the areas that needed to be considered was the effect of GFR-reinforced sand on symmetrical anchor plates. In this research, GFR-geogrid length (B') was not kept constant and the number of geogrid layers was varied for research. These components experienced elongation of a certain magnitude which is calculated in Appendix B. The variations of ACR with normalized layer spacing, $x/D = 0$ and 0.5 in various B'/D are illustrated in Figs. 4.17 and 4.18.

4.5.1 Test Series 5: Influence of Number and Vertical Spacing of GFR Layers on the Uplift Capacity

From Figs. 4.19 and 4.20, the overall trend indicates the effect of GFR, loosely and densely packed soil on symmetrical anchor plates with various numbers of geogrid layers and at an embedment depth between 1 and 4 layers. In reinforced tests, the geogrid layers were located at equal vertical spacings of $0.5D$, $0.75D$, and $1D$, with the first layer resting on the anchor plate. The variations in the symmetrical anchor plate's capacities with u/D

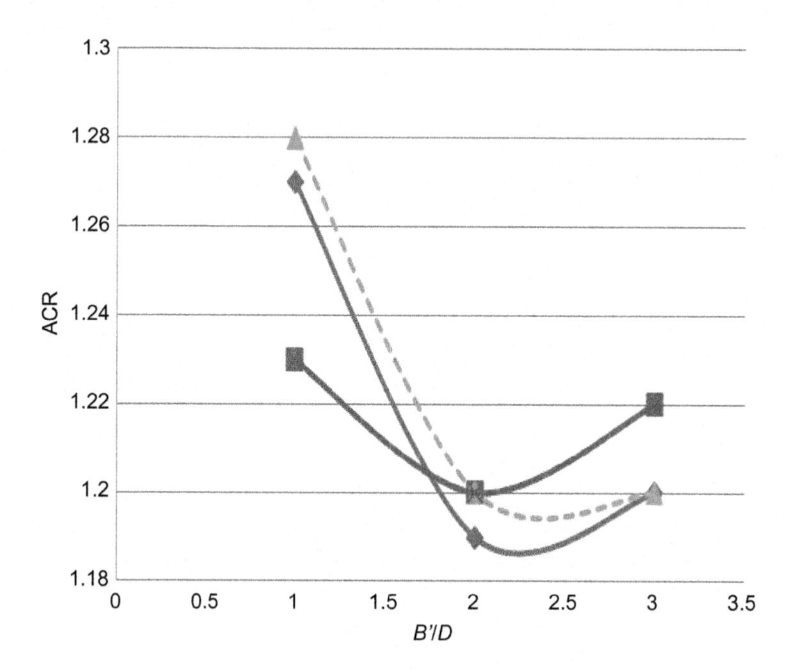

Figure 4.17 Variation of ACR with B'/D of GFR layer at $x/D = 0$.

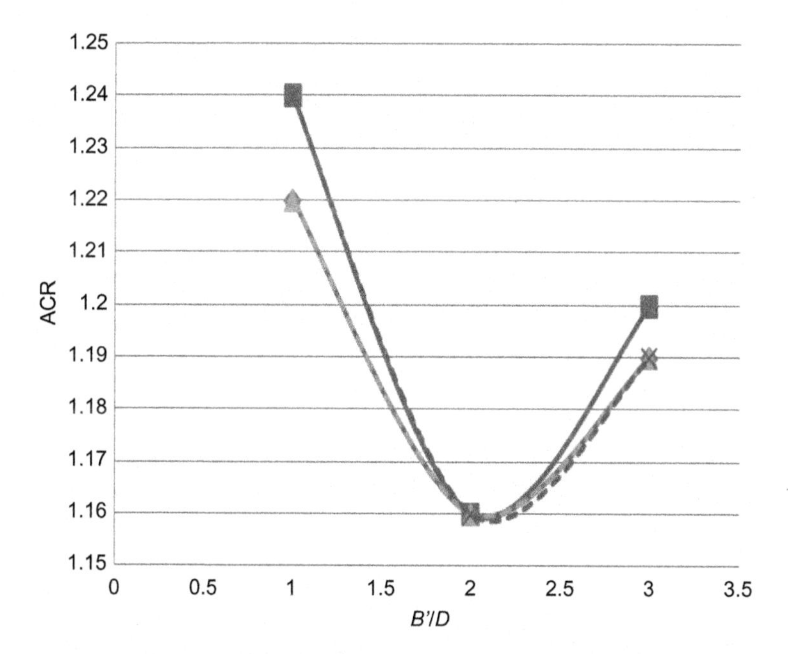

Figure 4.18 Variation of ACR with B'/D of GFR layer at $x/D = 0.5$.

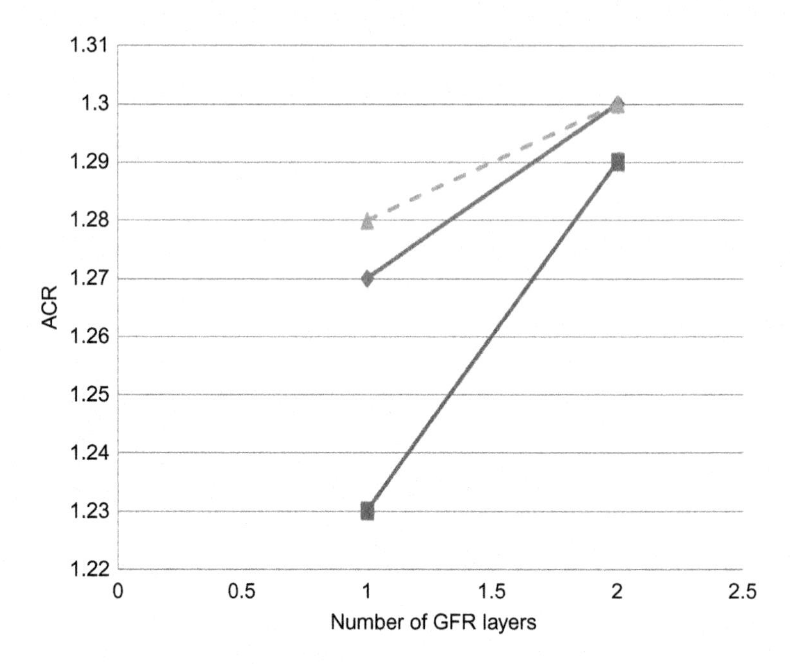

Figure 4.19 Variation of ACR with number of GFR layers at $x/D = 0$ and $u/D = 0.5$.

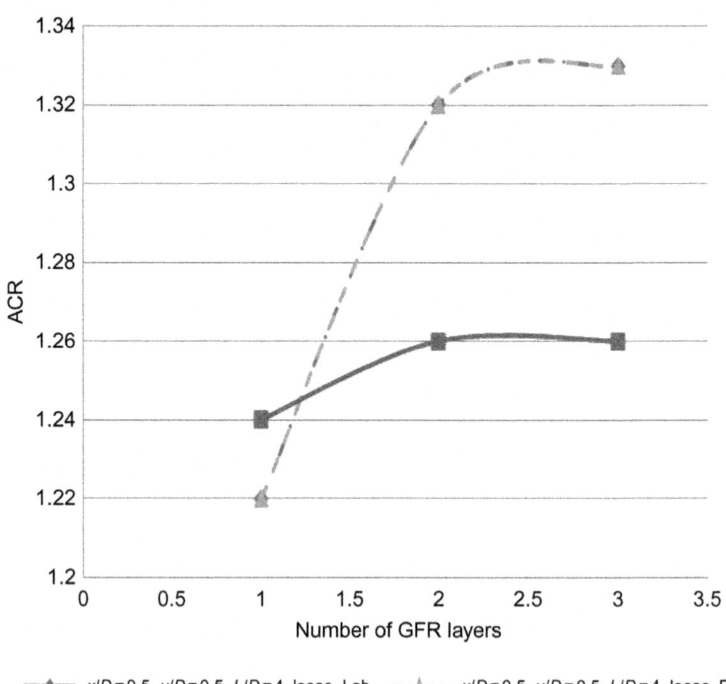

Figure 4.20 Variation of ACR with number of GFR layers at $x/D = 0.5$ and $u/D = 0.5$.

for a multilayers of GFR are plotted in Fig. 4.19. It can be seen that inclusion of GFR layers results in greater anchor capacity than in reinforced and non-reinforced sand. The conclusions are new for the uplift response of symmetrical anchor plates because earlier tests showed that a single layer of reinforcement close to the anchor plate was more effective than the use of multiple layers. However, GFR soil presented a significant effect of a single geogrid layer in the uplift response in both loosely and densely packed soil. Therefore, it was concluded that, in terms of symmetrical anchor plate's capacity, using a single layer of GFR is better and more effective than reinforcing the soil itself with multilayers. Hence, it was decided to carry out the test program on the response of anchor plates adjacent to loose or dense sand using a single layer of GFR placed in the symmetric state over the plate. This may be explained by the fact that the closeness of the GFR layer with adequate anchorage length to the anchor plate offers greater resistance to anchor displacement leading to greater failure wedge and larger failure surface area.

4.5.2 Test Series 6: Influence of GFR Layer Proximity to the Anchor on the Uplift Capacity

The overall trend indicates that for increasing uplift response, this part analyzed the influence of GFR layer proximity to the anchor on the uplift capacity. The data in Tables 4.6 and 4.7 indicate that many series of tests were performed on symmetrical anchor plates located in the loose and dense sand with the inclusion of one GFR layer placed at various distances of 0, $0.5D$, $0.75D$, and $1D$ over the symmetrical anchor plate. The data in Tables 4.6 and 4.7 indicate the variation of ACR response with x/D for the GFR-reinforced sand. Maximum gain in symmetrical anchor plate's capacity is attained when the GFR layer is placed resting directly on top of the anchor plate. This may be explained by the fact that the closeness of the GFR layer with adequate anchorage length to the anchor plate offers greater resistance to anchor displacement leading to greater failure wedge and larger failure surface area.

4.6 DISCUSSION ON UPLIFT CAPACITY OF SYMMETRICAL ANCHOR PLATES USING THE FINITE ELEMENT METHOD IN PLAXIS

Figs. 4.1–4.20 compare breakout factors and uplift responses from laboratory tests and PLAXIS finite element analyses for circular, square, and

Table 4.6 Variation of ACR with influence of GFR layer proximity to the anchor at $B'/D = 1$

Shapes	Embedment ratio	Test type	ACR based on GFR layer proximity to the symmetrical anchor plate in $B'/D = 1$							
	L/D	PLAXIS/laboratory	0		0.5		0.75		1	
Circular (10 cm)	4	Laboratory	1.21	1.21	1.19	1.20	1.18	1.20	1.17	1.18
Circular (10 cm)	4	PLAXIS	1.21	1.20	1.19	1.20	1.18	1.19	1.17	1.18
Square (10 cm)	4	Laboratory	1.23	1.27	1.24	1.22	1.23	1.24	1.23	1.23
Square (10 cm)	4	PLAXIS	1.23	1.28	1.24	1.22	1.23	1.24	1.23	1.23

Table 4.7 Variation of ACR with influence of GFR layer proximity to the anchor at $B'/D = 2$

Shapes	Embedment ratio	Test type	ACR based on GFR layer proximity to the symmetrical anchor plate in $B'/D = 2$							
	L/D	PLAXIS/laboratory	0		0.5		0.75		1	
Circular (10 cm)	4	Laboratory	1.15	1.15	1.14	1.14	1.16	1.15	1.18	1.18
Circular (10 cm)	4	PLAXIS	1.15	1.15	1.14	1.14	1.16	1.16	1.18	1.18
Square (10 cm)	4	Laboratory	1.20	1.19	1.16	1.16	1.19	1.19	1.23	1.22
Square (10 cm)	4	PLAXIS	1.20	1.20	1.16	1.16	1.19	1.19	1.23	1.23

rectangular anchor plates in dense and loose sand, respectively. In the case of densely packed soil, excellent agreement is found between the laboratory program and finite element modeling up to an embedment ratio of 4, although PLAXIS provided lower capacities than those derived from the laboratory program. A similar trend is evident for symmetrical anchor plates in loose packing conditions although the PLAXIS results are slightly higher than those observed for all embedded depths in this loose packing conditions.

4.7 OVERVIEW OF SOIL FAILURE MECHANISM STUDIES

Studies on the uplift failure mechanism have shown that symmetrical anchor plates have a curved shear surface failure, as illustrated in Figs. 4.21–4.28. The figures illustrate the shear failure mechanism during the uplift tests for symmetrical anchor plates in non-reinforced, reinforced and GFR loosely and densely packed soil. The condition of the soil surrounding the symmetrical anchor plates before the uplift test is shown in Fig. 4.21. Fig. 4.22 shows the failure pattern immediately after the uplift. Fig. 4.23 illustrates post-failure deformation experienced by the densely packed soil with subsequent uplift movement. A certain degree of collapse was observed near the symmetrical anchor plates. With further uplift movement, the failure surface was defined more prominently. The

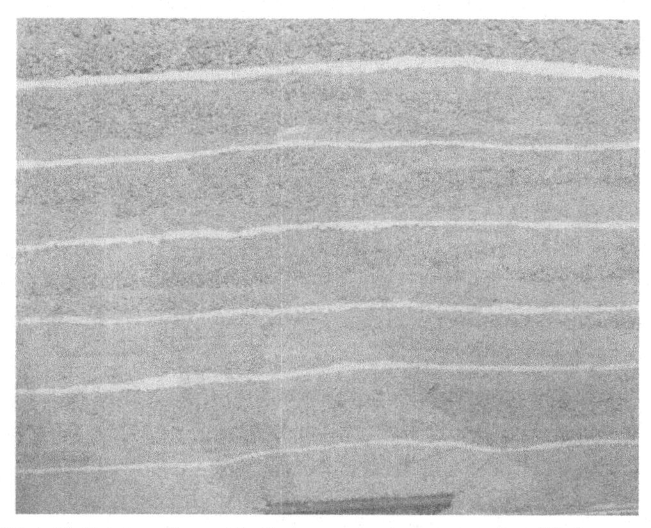

Figure 4.21 Initial state of sand before commencement of uplift in non-reinforced loose sand.

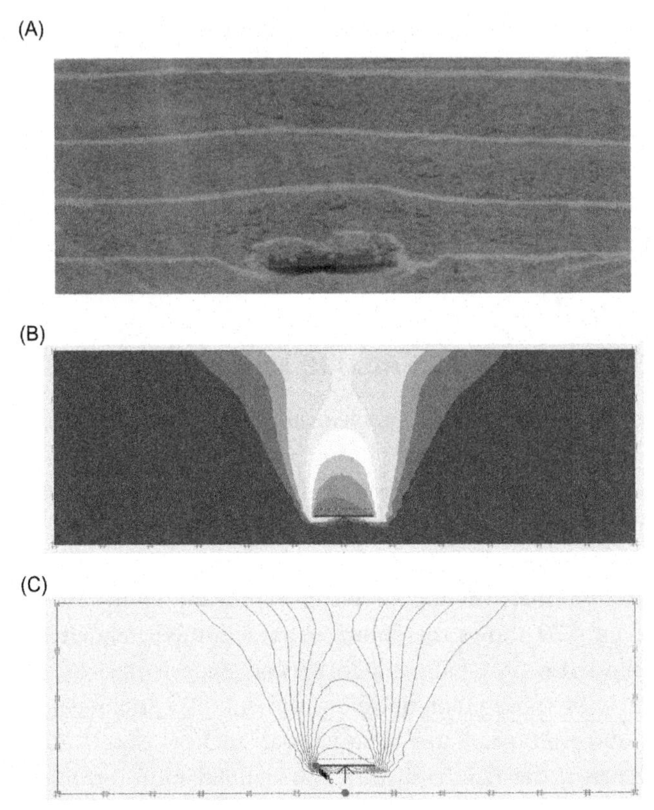

Figure 4.22 State of sand after commencement of uplift in non-reinforced loose sand. (A) Laboratory, (B) numerical simulation, (C) numerical simulation.

final failure surface is much more clearly shown in Fig. 4.23 where a curve-shaped localized failure was observed to have occurred when the symmetrical anchor plate was pulled out at a constant rate. A contributing factor towards the formation of the curved-shaped failure could be the collapse of the soil around the symmetrical anchor plate to fill in the void formed near the anchor plate's bottom. Figs. 4.24–4.28 show the behavior of the symmetrical anchor plate for different shapes during the uplift tests with reinforced and GFR conditions in loosely and densely packed soil during failure that indicates the formation of a shear zone along the uplift test. This is illustrated by the movement of soil particles along the symmetrical anchor-plate–soil interface that follows the symmetrical anchor plate during uplift. This shear zone comprising displaced soil particles along the symmetrical anchor-plate–soil interface is therefore seen

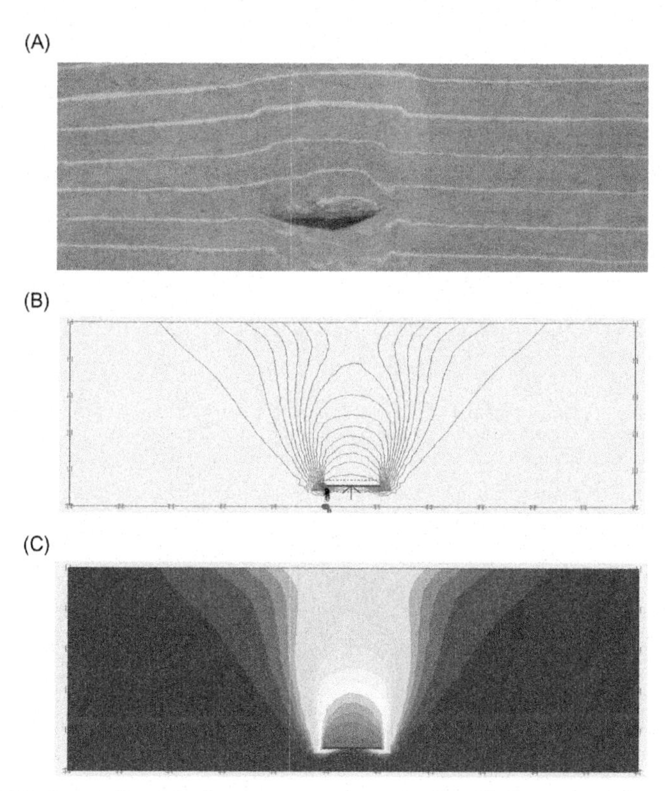

Figure 4.23 State of sand after commencement of uplift in non-reinforced dense sand. (A) Laboratory, (B) numerical simulation, (C) numerical simulation.

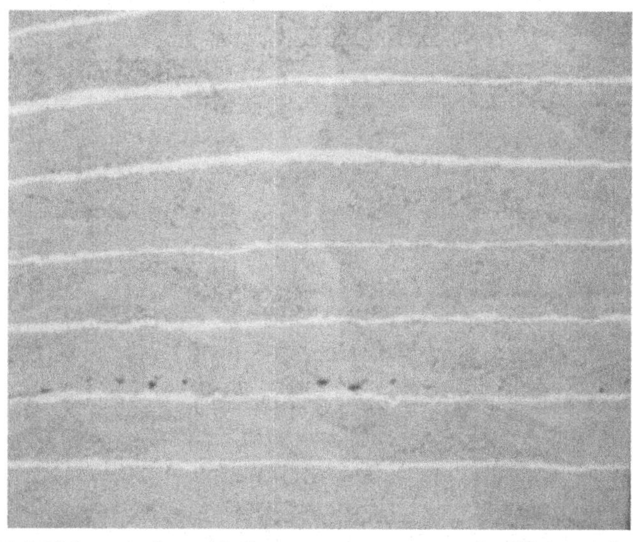

Figure 4.24 Initial state of sand before commencement of uplift in reinforced dense sand.

Figure 4.25 State of sand after commencement of uplift in reinforced dense sand.

to affect the increase in the uplift capacity. For symmetrical anchor plates with GFR, the development of the shear zone shown in Fig. 4.28 is restricted. The outfitted tension trend in the reinforcement allows the geogrid layer and GFR to resist the formed horizontal shear stresses built up in the soil mass inside the loaded zone and their movement beside the stable layers of loosely and densely packed soil, leading to a broader and deeper failure zone. Based on this result, soil–geogrid-layer interaction not only results in increasing the uplift capacity due to developed longer failure surface, but also results in extending the contact zone between the soil and the laboratory box.

4.8 EMPIRICAL RELATIONSHIP OF UPLIFT RESPONSE IN SAND

The data obtained from this research work will be used to develop an empirical relationship to compare the uplift response at the present embedment level with the existing theories and previous experimental works. The findings obtained will be used to derive empirical equations for symmetrical anchor plates, which will be based on the empirical

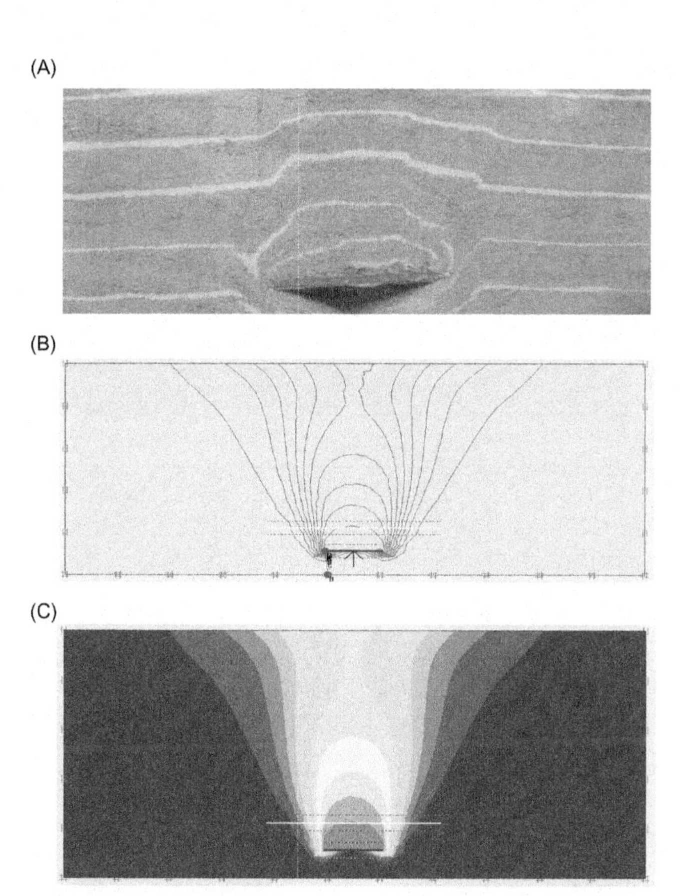

Figure 4.26 State of sand after failure of uplift in reinforced dense sand. (A) Laboratory, (B) numerical simulation, (C) numerical simulation.

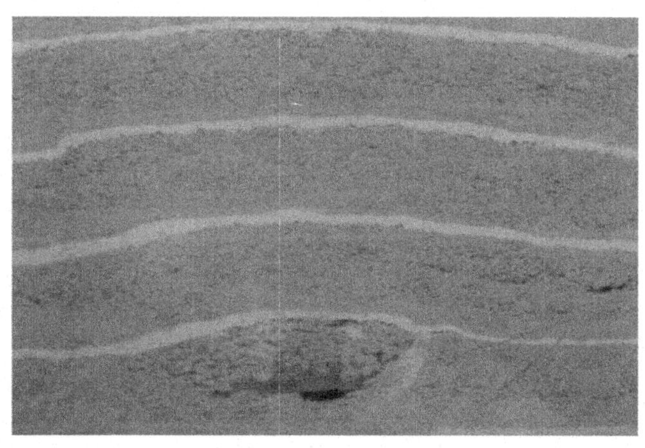

Figure 4.27 State of sand after commencement of uplift in GFR-reinforced dense sand.

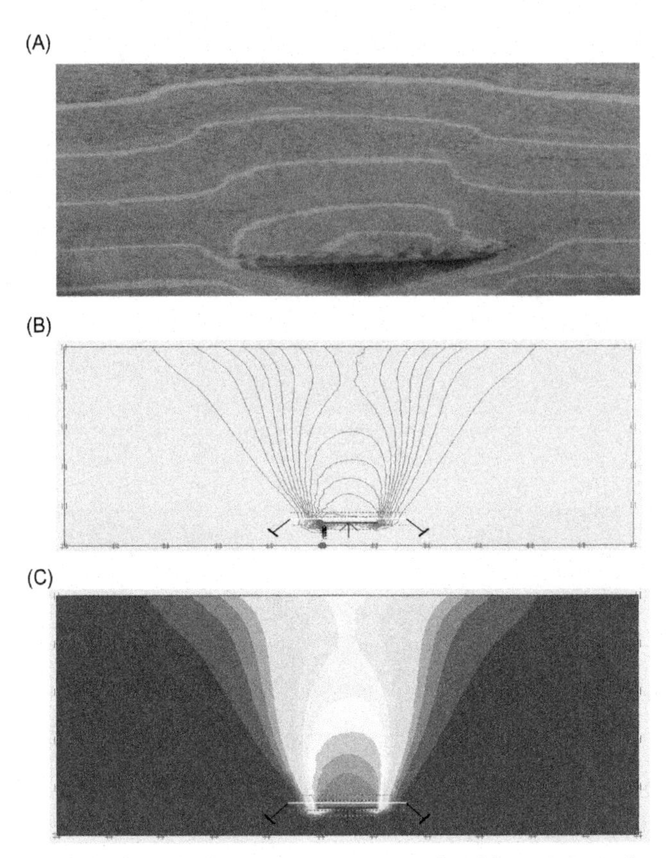

Figure 4.28 State of sand after commencement of uplift in GFR-reinforced dense sand. (A) Laboratory, (B) numerical simulation, (C) numerical simulation.

expression from the symmetrical anchor plates using the parameters mentioned in Chapter 3, Research Methodology. Comparison will be made between the uplift response and the breakout factor resulting from the calculations using the later equation with present test data and previous researcher's findings in Chapter 5, Comparison Between Existing Theories and Experimental Works.

4.8.1 Empirical Relationship for Non-reinforced Sand

Based on the results in this chapter, the variations of non-dimensional uplift responses with embedment ratio are plotted. Figs. 4.29, 4.31, and 4.33 derive the behavior of symmetrical anchor plates, including square, circular, and rectangular plates, in loose sand while Figs. 4.30, 4.32, and 4.34 show

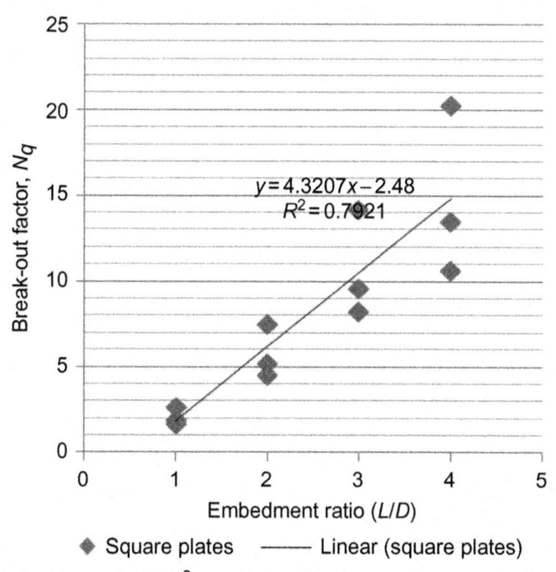

Figure 4.29 Variations of $P/LD^2\gamma$ with embedment ratio, L/D for square plate in loose sand.

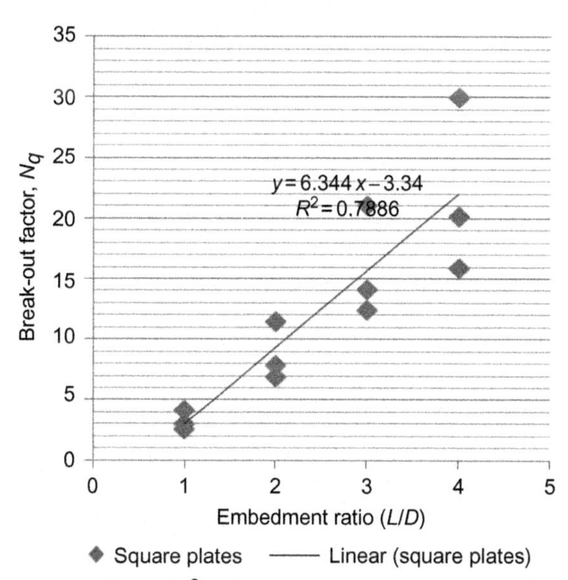

Figure 4.30 Variations of $P/LD^2\gamma$ with embedment ratio, L/D for square plate in dense sand.

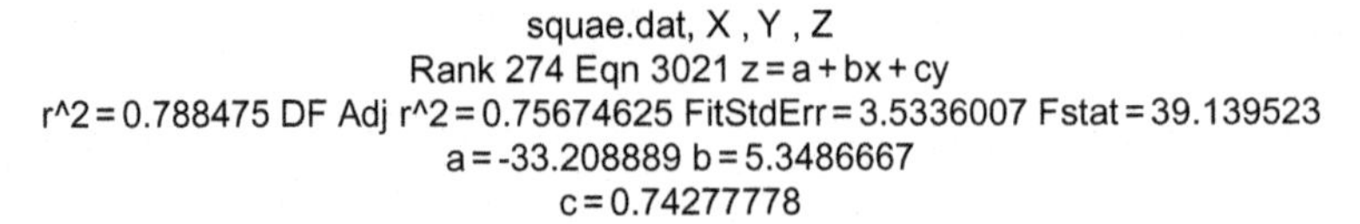

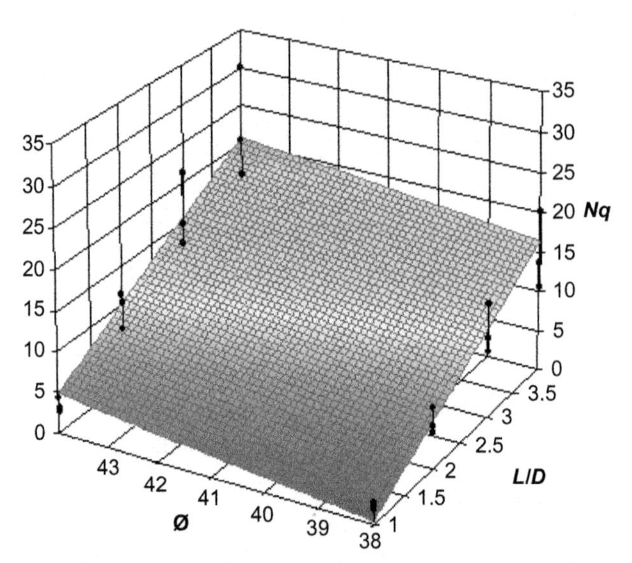

Figure 4.31 The general formula for square plates in non-reinforced sand.

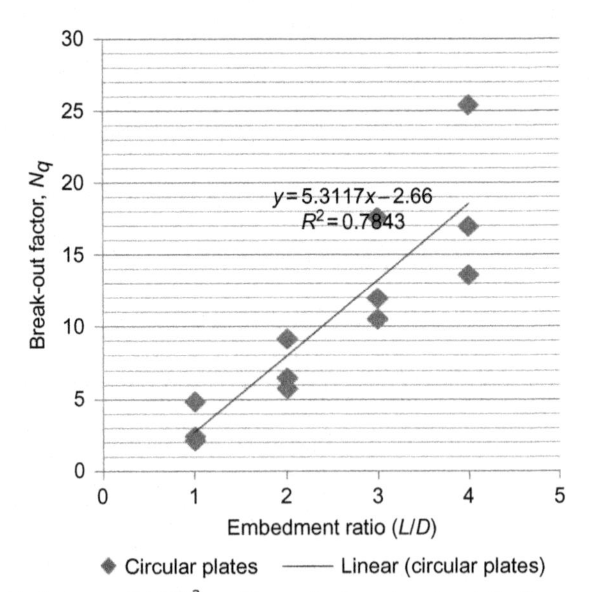

Figure 4.32 Variations of $P/LD^2\gamma$ with embedment ratio, L/D, for a circular plate in loose sand.

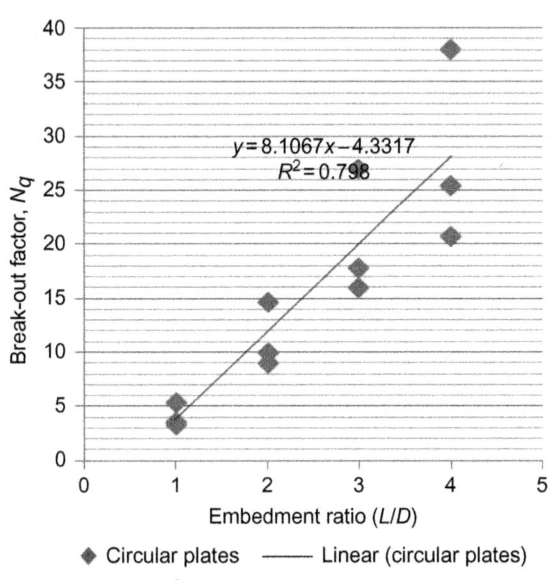

Figure 4.33 Variations of $P/LD^2\gamma$ with embedment ratio, L/D, for a circular plate in dense sand.

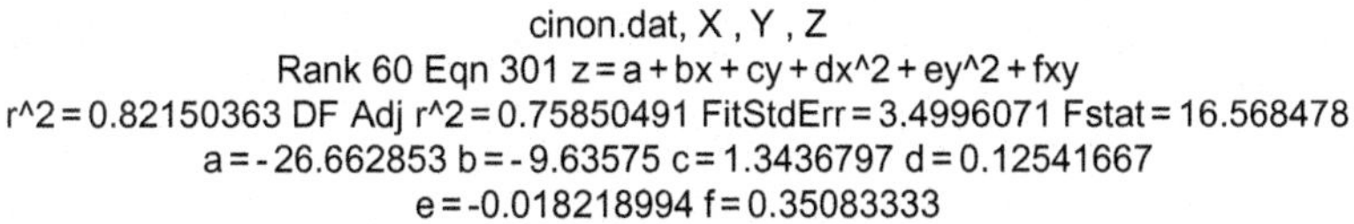

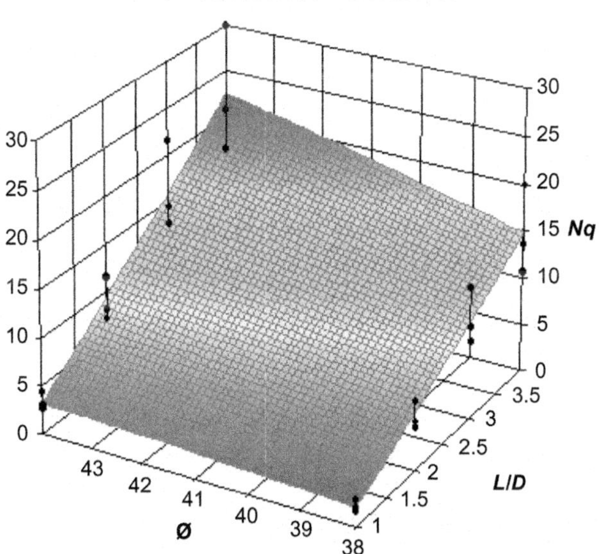

Figure 4.34 The general formula for circular plates in non-reinforced sand.

its behavior in dense packing. Tests in loose sand consist of three parts. The first part employed square anchor plates of various sizes, 50, 75, and 100 mm, as shown in Fig. 4.29; the second part employed circular anchor plates of various diameters, 50, 75, and 100 mm, as shown in Fig. 4.31; while the third part employed rectangular anchor plates of various dimensions, 200 and 300 mm, as shown in Fig. 4.33. Figs. 4.30, 4.32, and 4.33 show the similarity in symmetrical anchor plates in densely packed soil conditions. The empirical relationship for symmetrical anchor plates with sizes of 50, 75, and 100 mm for square and circular plates and 200 and 300 mm for rectangular plates in loose sand conditions were developed by combining both non-dimensional uplift responses to obtain average values. The methods were also adopted for dense conditions.

As shown in Figs. 4.29 and 4.30 the ratios of $P/\gamma LD^2$ were plotted against embedment ratio, L/D for symmetrical anchor plates. The empirical relationship was developed from test results. Linear regression was employed to obtain linear relationship of all data included. In square plates, this enables the following empirical relationship to be derived as:

$$N_q = 4.32(L/D) - 2.48 \tag{4.1}$$

Considering Fig. 4.30 for dense sand condition similar methods with loose sand were adopted. Thus empirical relationships for square anchor plates in dense sand condition gives:

$$N_q = 6.34(L/D) - 3.34 \tag{4.2}$$

As shown in Fig. 4.31 the general formula for square plates in non-reinforced sand was plotted.

In square plates, this enables the following empirical relationship as a new formula to be derived as:

$$P_{\text{square}} = \gamma D^2 L[-33.20 + 5.34(L/D) + 0.74(\varnothing)] \tag{4.3}$$

In circular plates, this enables the following empirical relationship to be derived as:

$$N_q = 5.13(L/D) - 2.66 \tag{4.4}$$

Considering Fig. 4.32 for dense sand conditions, similar methods to those for loose sand were adopted. Thus empirical relationships for circular anchor plates in dense sand conditions gives:

$$N_q = 8.10(L/D) - 4.33 \tag{4.5}$$

Figs. 4.32 and 4.33 show values of gradient plotted against embedment ratio, L/D giving a straight line by means of linear regression with coefficient of regression as Eqs. (4.4) and (4.5).

In circular plates, this enables the following empirical relationship as a new formula to be derived as:

$$P_{\text{circular}} = \gamma D^2 L[-26.66 + 9.63(L/D) + 1.34(\varnothing) + 0.12(L/D)^2 - 0.01(\varnothing)^2 + 0.35(L/D)(\varnothing)] \tag{4.6}$$

As shown in Fig. 4.34 the general formula for circular plates in non-reinforced sand was plotted.

In rectangular plates, this enables the following empirical relationship to be derived as:

$$N_q = 12.29(L/D) - 8.1 \tag{4.7}$$

Considering Fig. 4.36 for dense sand conditions, similar methods to those for loose sand were adopted. Thus empirical relationships for rectangular anchor plates in dense sand conditions give:

$$N_q = 18.21(L/D) - 11.48 \tag{4.8}$$

Figs. 4.35 and 4.36 show values of gradient plotted against embedment ratio, L/D, giving a straight line by means of linear regression with coefficient of regression as in Eqs. (4.7) and (4.8). As shown in Fig. 4.37, the general formula for rectangular plates in non-reinforced sand was plotted.

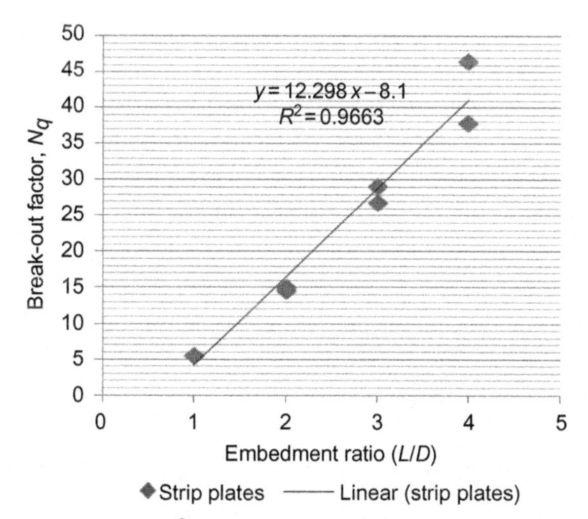

Figure 4.35 Variations of $P/LD^2\gamma$ with embedment ratio, L/D, for a rectangular plate in loose sand.

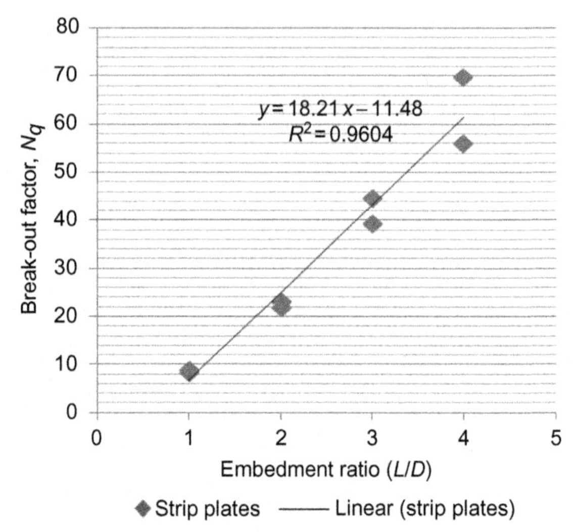

Figure 4.36 Variations of $P/LD^2\gamma$ with embedment ratio, L/D, for a rectangular plate in dense sand.

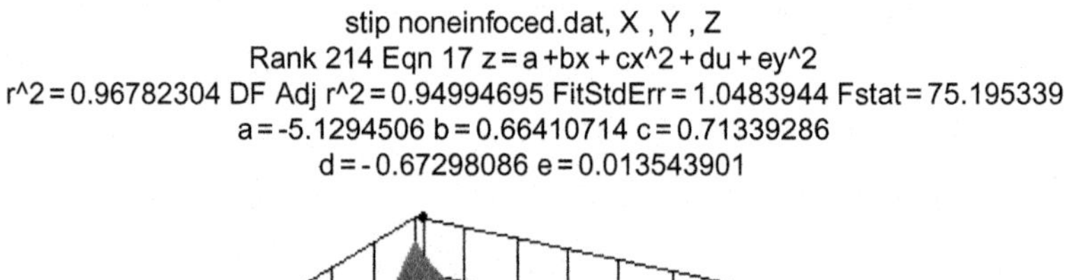

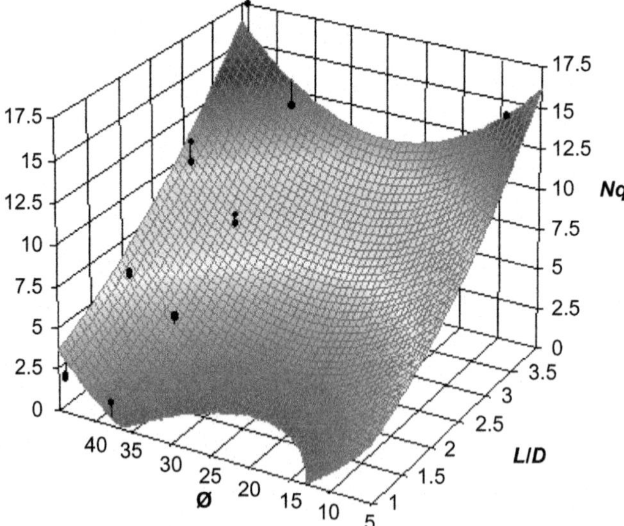

Figure 4.37 The general formula for rectangular plates in non-reinforced sand.

In rectangular plates, this enables the following empirical relationship as a new formula to be derived as:

$$P_{\text{circular}} = \gamma D^2 L\left[5.12 + 0.66(L/D) + 0.71(L/D)^2 - 0.67(\varnothing) + 0.01(\varnothing)^2\right] \tag{4.9}$$

4.8.2 Empirical Relationship for Reinforced Sand by Geogrid

Based on the results, the variations in non-dimensional uplift responses with embedment ratio are plotted. The results based on loose and dense condition derived the behavior of symmetrical anchor plates (square, circular, and rectangular) in reinforced sand. They are three series of tests, i.e., square, circular, and rectangular plates. For each series, the test was conducted for various sizes and embedment ratios. The first series employed square anchor plates with various numbers of geogrid layers and different x/D, B, and u/D. The second part employed circular anchor plates with various x/D, B, and u/D. The third part employed rectangular anchor plates with various numbers of geogrid layers and different x/D, B, and u/D. The following figures show the similarity in symmetrical anchor plates in densely packed soil conditions. The empirical relationship for symmetrical anchor plates with sizes of 50, 75, and 100 mm for square and circular plates and 200 and 300 mm for rectangular plates in loosely packed soil conditions using geogrid layers was developed by combining the non-dimensional uplift responses to obtain the average values. The analysis examined these two hypotheses:

Hypothesis 1: the proposed geogrid layer(s) could improve the breakout factor of symmetrical anchor plates in loose sand; and

Hypothesis 2: The proposed geogrid layer(s) could improve the breakout factor of symmetrical anchor plates in dense sand.

The analysis measured the breakout factor of symmetrical anchor plates in reinforced soil. In the research assigned several items to each of the dimensions. The five dimensions are independent variables and the breakout factor is a dependent variable in the multiple regression analysis. This analysis predicts the dependent variable using the independent variables. As there are five independent variables in this analysis, the regression equation of the analysis is:

$$Y = \beta_0 + \beta_1 X_1 + \beta_2 X_2 + \beta_3 X_3 + \beta_4 X_4 + \beta_5 X_5$$

where, Y, dependent variable (predicted by regression model); X_i ($i=1,2,\ldots,5$), ith independent variable from total set of k variables; β_i ($i=1,2,\ldots,5$), ith coefficient corresponding to X_i; β_0, intercept (or constant); $i=1,2,\ldots,5$, independent variables' index.

The regression analysis determines the impact of each of these independent variables on the dependent variable by their coefficients, so we have the following hypotheses for each of these independent variables:

H_0 (null hypothesis): $\beta_i = 0$, the independent variable, X_i, is not important for predicting the dependent variable (Y); and

H_1 (alternative hypothesis): $\beta_i \neq 0$, the independent variable, X_i, is important for predicting the dependent variable (Y).

The coefficients show the impact of these independent variables on the dependent variable and the signs of these coefficients show the direction of the impacts (positive or negative). The P-value proves or rejects the null hypothesis. If the P-value of X_i is less than the level of significance (0.05), then the null hypothesis is rejected, but if the P-value of X_i is greater than the level of significance (0.05), then the null hypothesis is proved.

As shown in Tables 4.8–4.11, the relationships between various parameters were plotted for symmetrical anchor plates. The empirical relationship was developed from tests results. Linear regression was employed to obtain a linear relationship of all data included. In square plates, this enables the following empirical relationship to be derived as:

$$N_q = 1.17 + 2.88(L/D) - 0.12(N) \tag{4.10}$$

Considering Table 4.9 for dense sand conditions, similar methods to those in loose sand were adopted. Thus empirical relationships for square anchor plates in dense sand conditions gives:

$$N_q = 2.23 + 4.24(L/D) - 0.14(N) \tag{4.11}$$

Tables 4.8 and 4.9 show formulas plotted against embedment ratio, L/D, and number of geogrid layers (N) giving a linear equation by means of linear regression with coefficient of regression as Eqs. (4.10) and (4.11).

As the confidence level has been set up on 95% in the analysis, P-value of significant variables should be less than 0.05. According to Tables 4.8 and 4.9, "embedment ratio" and "number of geogrid layers" have significant impacts on the overall service quality. The impact of "embedment ratio" is positive and the impact of "number of geogrid layers" is negative.

Table 4.8 Multiple regression analysis results between the overall data of square plates and the five factor dimensions in loose sand using geogrid

Regression statistics

Multiple R	0.997149
R^2	0.994306
Adjusted R^2	0.993357
Standard error	0.119783

ANOVA

	df	*SS*	*MS*	*F*	Significance *F*
Regression	5	75.16235	15.03247	1047.708	1.05E-32
Residual	30	0.430439	0.014348		
Total	35	75.59279			

	Coefficients	Standard error	*t* Stat	*P*-value	Lower 95%
Intercept	1.175358	0.162793	7.219938	4.9E-08	0.842889
X Variable 1	2.883333	0.039928	72.21394	3.32E-35	2.80179
X Variable 2	− 0.12522	0.062062	− 2.01773	0.052642	− 0.25197
X Variable 3	− 0.04244	0.065081	− 0.65204	0.519341	− 0.17535
X Variable 4	− 0.01754	0.189805	− 0.09239	0.927003	− 0.40517
X Variable 5	0.045984	0.027225	1.689011	0.101587	− 0.00962

Table 4.9 Multiple regression analysis results between the overall data of square plates and the five factor dimensions in dense sand using geogrid

Regression statistics

Multiple R	0.994402
R^2	0.988835
Adjusted R^2	0.986974
Standard error	0.247046

ANOVA

	df	*SS*	*MS*	*F*	Significance *F*
Regression	5	162.1603	32.43207	531.3951	2.53E-28
Residual	30	1.830958	0.061032		
Total	35	163.9913			

	Coefficients	Standard error	*t* Stat	*P*-value	Lower 95%
Intercept	2.239141	0.335753	6.66901	2.19E-07	1.553442
X Variable 1	4.241111	0.082349	51.50179	7.78E-31	4.072932
X Variable 2	− 0.14205	0.128	− 1.1098	0.275904	− 0.40347
X Variable 3	0.02155	0.134227	0.160548	0.873526	− 0.25258
X Variable 4	0.103583	0.391463	0.264606	0.793122	− 0.69589
X Variable 5	0.060283	0.056151	1.073584	0.291569	− 0.05439

Table 4.10 Multiple regression analysis results between the overall data of circular plates and the five factor dimensions in loose sand using geogrid

Regression statistics

Multiple R	0.997495
R^2	0.994995
Adjusted R^2	0.994161
Standard error	0.108686

ANOVA

	df	*SS*	*MS*	*F*	Significance *F*
Regression	5	70.4563	14.09126	1192.902	1.51E-33
Residual	30	0.354378	0.011813		
Total	35	70.81068			

	Coefficients	Standard error	*t* Stat	*P*-value	Lower 95%
Intercept	1.369444	0.147711	9.27108	2.59E-10	1.067777
X Variable 1	2.793889	0.036229	77.11836	4.68E-36	2.7199
X Variable 2	− 0.12583	0.056312	− 2.23456	0.033046	− 0.24084
X Variable 3	− 0.04	0.059052	− 0.67737	0.503361	− 0.1606
X Variable 4	0.152222	0.17222	0.883881	0.383789	− 0.1995
X Variable 5	0.063333	0.024703	2.563778	0.015606	0.012883

Table 4.11 Multiple regression analysis results between the overall data of circular plates and the five factor dimensions in dense sand using geogrid

Regression statistics

Multiple R	0.995373
R^2	0.990768
Adjusted R^2	0.98923
Standard error	0.221171

ANOVA

	df	*SS*	*MS*	*F*	Significance *F*
Regression	5	157.4953	31.49907	643.9323	1.47E-29
Residual	30	1.467502	0.048917		
Total	35	158.9628			

	Coefficients	Standard error	*t* Stat	*P*-value	Lower 95%
Intercept	2.37832	0.300587	7.912251	7.87E-09	1.764439
X Variable 1	4.180556	0.073724	56.70567	4.46E-32	4.029992
X Variable 2	− 0.13927	0.114594	− 1.21532	0.233722	− 0.3733
X Variable 3	0.010406	0.120168	0.086595	0.931569	− 0.23501
X Variable 4	0.207552	0.350462	0.592225	0.558137	− 0.50819
X Variable 5	0.06928	0.05027	1.378169	0.178344	− 0.03338

In circular plates, this enables the following empirical relationship to be derived as:

$$N_q = 1.36 + 2.79(L/D) - 0.12(N) + 0.06(B/D) \tag{4.12}$$

Considering Table 4.11 for dense sand condition similar methods with loose sand were adopted. Thus empirical relationships for circular anchor plates in dense sand condition gives:

$$N_q = 2.23 + 4.24(L/D) - 0.14(N) \tag{4.13}$$

Tables 4.10 and 4.11 formulas plotted against embedment ratio, L/D and number of geogrid layers (N) giving a linear equation by means of linear regression with coefficient of regression as Eqs. (4.12) and (4.13).

As the confidence level has been set up on 95% in the analysis, P-value of significant variables should be less than 0.05. According to Tables 4.10 and 4.11, "embedment ratio," "number of geogrid layers," and "ratio of geogrid width to plate width" have significant impacts on the overall service quality. The impact of "embedment ratio" and "ratio of geogrid width to plate width" are positive, and the impact of "number of geogrid layers" is negative.

4.8.3 Empirical Relationship for Reinforced Sand by GFR

Based on the results in this chapter, the variations of non-dimensional uplift responses with embedment ratio are plotted. The results based on loose and dense conditions derived the behavior of symmetrical anchor plates (square, circular, and rectangular plates) in sand using GFR. Tests in loose sand conditions consist of three parts. The first part employed square plates with sizes of 50, 75, and 100 mm, and various numbers of GFR layers and various x/D, B', and u/D. The second part employed circular plates of diameters of 50, 75, and 100 mm, and various numbers of GFR layers and various x/D, B', and u/D. The third part employed rectangular plates with of diameters 200 and 300 mm and various numbers of GFR layers and various x/D, B', and u/D. The empirical relationship for symmetrical anchor plates with sizes of 50, 75, and 100 mm for square and circular plates, and 200 and 300 mm for rectangular plates in loose sand conditions using GFR were developed by combining both non dimensional uplift responses to obtain average values. The methods were also adopted for dense conditions.

As shown in Tables 4.12–4.15, the relationships between various parameters were plotted for symmetrical anchor plates. The empirical relationship was developed from tests results. Linear regression was employed to obtain linear relationships of all data included. In square plates in loose sand, this enables the following empirical relationship to be derived as:

$$N_q = 1.98 + 3(L/D) - 0.56(B'/D) \tag{4.14}$$

Considering Table 4.13 for dense sand conditions, similar methods to those for loose sand were adopted. Thus empirical relationships for square anchor plates in dense sand conditions gives:

$$N_q = 6.57 + 3.45(L/D) - 0.49(B'/D) \tag{4.15}$$

Tables 4.12 and 4.13 show formulas plotted against embedment ratio, *L/D*, and number of GFR layers (*N*) giving a linear equation by means of linear regression with coefficient of regression as in Eqs. (4.14) and (4.15).

As the confidence level has been set up as 95% in the analysis, the *P*-value of significant variables should be less than 0.05. According to Tables 4.12 and 4.13, "embedment ratio" and "ratio of geogrid width to plate width" have significant impacts on the overall service quality. The

Table 4.12 Multiple regression analysis results between the overall data of square plates and the five factor dimensions in loose sand using GFR

Regression statistics

Multiple *R*	0.976997
R^2	0.954523
Adjusted R^2	0.946943
Standard error	0.372875

ANOVA

	df	*SS*	*MS*	*F*	Significance *F*
Regression	5	87.54676	17.50935	125.9343	3.4E-19
Residual	30	4.171068	0.139036		
Total	35	91.71783			

	Coefficients	Standard error	*t* Stat	*P*-value	Lower 95%
Intercept	1.980885	0.506763	3.908901	0.00049	0.945937
X Variable 1	3.003889	0.124292	24.16808	3.19E-21	2.750052
X Variable 2	0.027442	0.193195	0.142041	0.887997	−0.36711
X Variable 3	−0.2831	0.202592	−1.39738	0.172548	−0.69685
X Variable 4	0.508389	0.590847	0.860441	0.39637	−0.69828
X Variable 5	−0.56223	0.084751	−6.63397	2.41E-07	−0.73532

Table 4.13 Multiple regression analysis results between the overall data of square plates and the five factors dimensions in dense sand using GFR

Regression statistics

Multiple R	0.975685
R^2	0.951962
Adjusted R^2	0.943955
Standard error	0.435005

ANOVA

	df	*SS*	*MS*	*F*	Significance *F*
Regression	5	112.497	22.4994	118.9003	7.7E-19
Residual	30	5.676873	0.189229		
Total	35	118.1739			

	Coefficients	Standard error	*t* Stat	*P*-value	Lower 95%
Intercept	6.571443	0.591201	11.11541	3.69E-12	5.364049
X Variable 1	3.451667	0.145002	23.80434	4.91E-21	3.155534
X Variable 2	− 0.03061	0.225385	− 0.13582	0.892871	− 0.49091
X Variable 3	− 0.26089	0.236349	− 1.10381	0.278449	− 0.74358
X Variable 4	0.824936	0.689296	1.196781	0.240764	− 0.58279
X Variable 5	− 0.4904	0.098872	− 4.95995	2.61E-05	− 0.69232

Table 4.14 Multiple regression analysis results between the overall data of circular plates and the five factor dimensions in loose sand using GFR

Regression statistics

Multiple R	0.97415
R^2	0.948968
Adjusted R^2	0.940463
Standard error	0.377633

ANOVA

	df	*SS*	*MS*	*F*	Significance *F*
Regression	5	79.55509	15.91102	111.573	1.9E-18
Residual	30	4.27819	0.142606		
Total	35	83.83328			

	Coefficients	Standard error	*t* Stat	*P*-value	Lower 95%
Intercept	2.295651	0.513229	4.472959	0.000103	1.247498
X Variable 1	2.855	0.125878	22.68078	1.95E-20	2.597924
X Variable 2	0.029508	0.19566	0.150813	0.881133	− 0.37008
X Variable 3	− 0.29137	0.205178	− 1.42006	0.165898	− 0.71039
X Variable 4	0.774092	0.598386	1.293634	0.205664	− 0.44797
X Variable 5	− 0.52652	0.085832	− 6.1343	9.58E-07	− 0.70181

Table 4.15 Multiple regression analysis results between the overall data of circular plates and the five factor dimensions in dense sand using GFR

Regression statistics

Multiple R	0.975742
R^2	0.952072
Adjusted R^2	0.944084
Standard error	0.417373

ANOVA

	df	*SS*	*MS*	*F*	Significance *F*
Regression	5	103.8133	20.76266	119.1886	7.44E-19
Residual	30	5.226003	0.1742		
Total	35	109.0393			

	Coefficients	Standard error	*t* Stat	*P*-value	Lower 95%
Intercept	6.914943	0.567238	12.19054	3.76E-13	5.756487
X Variable 1	3.302778	0.139124	23.73976	5.31E-21	3.018648
X Variable 2	− 0.0055	0.21625	− 0.02544	0.979873	− 0.44714
X Variable 3	− 0.36133	0.226769	− 1.59337	0.12156	− 0.82445
X Variable 4	1.000738	0.661357	1.513159	0.140704	− 0.34993
X Variable 5	− 0.4581	0.094864	− 4.82899	3.77E-05	− 0.65184

impact of "embedment ratio" is positive and the impact of and "ratio of geogrid width to plate width" is negative.

In circular plates, this enables the following empirical relationship to be derived as:

$$N_q = 2.29 + 2.85(L/D) - 0.52(B'/D) \tag{4.16}$$

Considering Table 4.15 for dense sand conditions, similar methods to those for loose sand were adopted. Thus empirical relationships for square anchor plates in dense sand conditions gives:

$$N_q = 6.91 + 3.30(L/D) - 0.45(B'/D) \tag{4.17}$$

Tables 4.14 and 4.15 show formulas plotted against embedment ratio, L/D, and number of GFR layers (N) giving a linear equation by means of linear regression with coefficient of regression as in Eqs. (4.16) and (4.17).

As the confidence level has been set up as 95% in the analysis, the P-value of significant variables should be less than 0.05. According to Tables 4.14 and 4.15, "embedment ratio" and "ratio of geogrid width to plate width" have significant impacts on the overall service quality. The impact of "embedment ratio" is positive and the impact of "ratio of geogrid width to plate width" is negative.

The equations are valid for symmetrical anchor plates in loose and dense sand conditions with a restriction of $L/D \leq 4$.

Tables 4.16–4.21 show the interpretations of the results from Eqs. (4.1) to (4.17). Tables 4.16 and 4.17 illustrate comparison uplift

Table 4.16 Comparison between prediction and experimental uplift load for symmetrical anchor plates in embedment ratio, $L/D = 4$, in loose sand under non-reinforced conditions

Plates	Size (mm)	PUR	EUR	% Diff
Square	50	111	152	− 26.97
Square	75	374.58	342	9.5
Square	100	888	640	38.75
Circular	50	139.35	150	− 7.1
Circular	75	470.25	339	38.71
Circular	100	1114.8	644	73.10

Table 4.17 Comparison between prediction and experimental uplift load for symmetrical anchor plates in embedment ratio, $L/D = 4$, in dense sand under non-reinforced conditions

Plates	Size (mm)	PUR	EUR	% Diff
Square	50	187.17	256	− 26.88
Square	75	631.53	579	9.07
Square	100	1497	1088	37.59
Circular	50	238.59	254	− 6.06
Circular	75	805.04	576	39.76
Circular	100	1908.76	1108	72.27

Table 4.18 Comparison between prediction and experimental uplift load for symmetrical square plates in embedment ratio, $L/D = 4$, $x/D = 0.5$, in loose sand under geogrid-reinforced conditions

N	*B/D*	PUR	EUR	% Diff
1	1	754.2	777	− 2.9
1	2	754.2	752	0.29
1	3	754.2	760	− 0.76
2	1	747	741	0.80
2	2	747	751	− 0.53
2	3	747	754	− 0.92
3	1	739.8	741	− 0.16
3	2	739.8	741	− 0.16
3	3	739.8	737	0.37

Table 4.19 Comparison between prediction and experimental uplift load for symmetrical square plates in embedment ratio, $L/D=4$, $x/D=0.5$, in dense sand under geogrid-reinforced conditions

N	*B/D*	PUR	EUR	% Diff
1	1	1345.72	1299	3.5
1	2	1345.72	1261	6.7
1	3	1345.72	1295	3.9
2	1	1336.2	1281	4.3
2	2	1336.2	1302	2.6
2	3	1336.2	1299	2.8
3	1	1326.68	1286	3.1
3	2	1326.68	1289	2.9
3	3	1326.68	1270	4.4

Table 4.20 Comparison between prediction and experimental uplift load for symmetrical square plates in embedment ratio, $L/D=4$, $x/D=0.5$, in loose sand under GFR-reinforced conditions

N	*B/D*	PUR	EUR	% Diff
1	1	804	786	2.29
1	2	769.2	744	3.38
1	3	734.4	765	−4
2	1	804	847	−5.07
2	2	769.2	756	−1.74
2	3	734.4	757	−2.98
3	1	804	856	−6.07
3	2	769.2	753	2.15
3	3	734.4	735	−0.08

Table 4.21 Comparison between prediction and experimental uplift load for symmetrical square plates in embedment ratio, $L/D=4$, $x/D=0.5$, in dense sand under GFR-reinforced conditions

N	*B/D*	PUR	EUR	% Diff
1	1	1351.84	1356	−0.30
1	2	1318.52	1264	4.31
1	3	1285.2	1306	−1.59
2	1	1351.84	1379	−1.96
2	2	1318.52	1308	0.80
2	3	1285.2	1310	−1.89
3	1	1351.84	1381	−2.10
3	2	1318.52	1322	−0.26
3	3	1285.2	1269	1.27

empirical expression and empirical expression for symmetrical anchor plates (50, 75, and 100 mm) in loose and dense conditions based on nonreinforced sand, while Tables 4.18–4.22 show symmetrical anchor plates of 100 mm size in sand using geogrid and GFR. These tables consist of predictions uplift response (PUR), experimental uplift response (EUR), and percentage difference (% Diff.). The negative signs show that the prediction uplift response (PUR) are underestimated, while the positive signs show that it is overestimated. In other words, if uplift response predicted is less than EUR, the value is said to be underestimated.

4.9 SUMMARY OF RESULTS

From the detailed analysis given above, the findings of the parametric research can be summarized. Based on the experimental and numerical studies carried out on symmetrical anchor plates (e.g., square, circular, and rectangular anchor plates) that were embedded adjacent to an experimental box at two sand densities with and without geogrid reinforcement and GFR, the following conclusions are drawn:

- Inclusion of geogrid reinforcement in a laboratory test chamber significantly increases the ultimate pullout resistance of a symmetrical anchor plate embedded in sand.
- In cases where design requirements necessitate large uplift resistance, soil reinforcement could be considered as an economical solution and could be used to obtain the designed symmetrical anchor plate capacity instead of increasing the embedment depth or anchor size.
- In terms of anchor capacity, inclusion of one layer of geogrid over the anchor plate is more cost effective than sand reinforcement using multiple layers. The optimal location of one geogrid inclusion is when it is resting directly on top of the symmetrical anchor plate.
- In terms of symmetrical anchor capacity, inclusion of multilayers of geogrid over the anchor plate is not more effective than soil reinforcement using one layer. In terms of reinforced conditions on a symmetrical anchor plate using multilayers of geogrid, the optimal space of multigeogrid layer inclusion is 0.5*D*. For practical reasons, no tests were carried out using more than three geogrid layers due to the limited depth of the sand fill. However, the result indicates that for the certain sand and reinforcement conditions, there are a critical number

of reinforcement layers after which the improvement in uplift capacity is not only inconsiderable but also has an opposite effect. This increase in uplift capacity of the anchor plate can be related to the reinforcement mechanism which is formed from the passive earth resistance, engaging in front of the transverse members, and adhesion between the longitudinal/transverse geogrid members and the sand. The mobilized tension in the reinforcement allows the geogrid to resist the formed horizontal shear stresses built up in the soil mass inside the loaded area and transfers them to the stable layers of soil beside the anchor plate, leading to a broader and deeper failure zone.

- Increased soil density and embedment depth results in greater uplift capacity.
- Inclusion of GFR in an experimental box significantly increases the ultimate uplift response of a symmetrical anchor plate embedded in sand. However, GFR is much better than geogrid in improving the symmetrical anchor plate's capacity. This is because of the presence of anchors in GFR which provide higher pullout strength. However, the intrinsic merit of GFR is brought about by its anchorage strength or pullout resistance, which can far exceed the direct shear strength. The mobilized tension in the reinforcement allows the GFR to resist the formed horizontal shear stresses built up in the soil mass inside the loaded area, and transfers them to the stable layers of soil beside the anchor plate, leading to a broader and deeper failure zone.
- In terms of anchor capacity, inclusion of multilayers of GFR over the anchor plate is not more effective than soil reinforcement using one layer. The optimal space of multi-GFR layer inclusion is when it is $0.5D$.
- GFR increases the ultimate uplift response of a symmetrical anchor plate embedded in sand.
- GFR increased the uplift capacity for symmetrical anchor plates in both loosely and densely packed sand.
- Inclusion of geogrid reinforcement in a laboratory chamber significantly increases the uplift force due to the development of a longer failure surface, but also results in an extension of the contact zone between the soil and the laboratory box.
- Inclusion of GFR in an experimental box significantly increases the uplift force due to the development of a longer failure surface, but also results in an extension of the contact zone between the soil and the laboratory box.

REFERENCES

Balla, A. (1961). The resistance of breaking-out of mushroom foundations for pylons. *Proceedings of the Fifth International Conference on Soil Mechanics and Foundation Engineering*, *1*, 569–576.

Meyerhof, G. G., & Adams, J. I. (1968). The ultimate uplift capacity of foundations. *Canadian Geotechnical Journal*, *5*(4), 225–244.

Vesic, A. S. (1971). Breakout resistance of objects embedded in ocean bottom. *Journal of the Soil Mechanics and Foundations Division ASCE*, *97*(9), 1183–1205.

CHAPTER 5

Comparison Between Existing Theories and Experimental Works

5.1 INTRODUCTION

This chapter presents a comparison of theoretical and experimental values for the experimental and numerical programs conducted. Chapter 2, Literature Review, explains previous theoretical research results, which were dedicated to the limiting ultimate uplift capacity of symmetrical anchor plates, their breakout factor, and failure zones. Researchers such as Balla (1961), Meyerhof and Adams (1968), Vesic (1971), Rowe and Davis (1982), Murray and Geddes (1987), Sarac (1989), Smith (1998), Krishna (2000), Frgic and Marovic (2003), Merifield and Sloan (2006), Dickin and Laman (2007), Kumar and Bhoi (2008), and Kuzer and Kumar (2009) determined the parametric relationship between ultimate uplift capacity of anchor plates and their breakout factor.

Researchers like, Balla (1961), Meyerhof and Adams (1968), Vesic (1971), Rowe and Davis (1982), Murray and Geddes (1987), dedicated their works to proposing the theories of horizontal anchor plates subjected to uplift loads. This chapter presents a comparison of the author's findings with existing theories in current research.

5.2 COMPARISON OF THE BREAKOUT FACTOR BETWEEN CURRENT RESULTS AND EXPERIMENTAL AND PREDICTED VALUES

Figs. 5.1–5.6 illustrate comparisons of theoretical and experimental values as proposed by various researchers such as Balla (1961), Meyerhof and Adams (1968), Vesic (1971), Rowe and Davis (1982), and Murray and Geddes (1987). These authors proposed theoretical values based on the curved failure model by the method of analytical and experimental evaluation in non-reinforced sand. The following sections are a summary of breakout factors for uplift capacity used as a basis for comparison with theoretical and experimental results.

Soil Reinforcement for Anchor Plates and Uplift Response.
DOI: http://dx.doi.org/10.1016/B978-0-12-809558-4.00005-4
© 2017 Elsevier Inc.
All rights reserved.

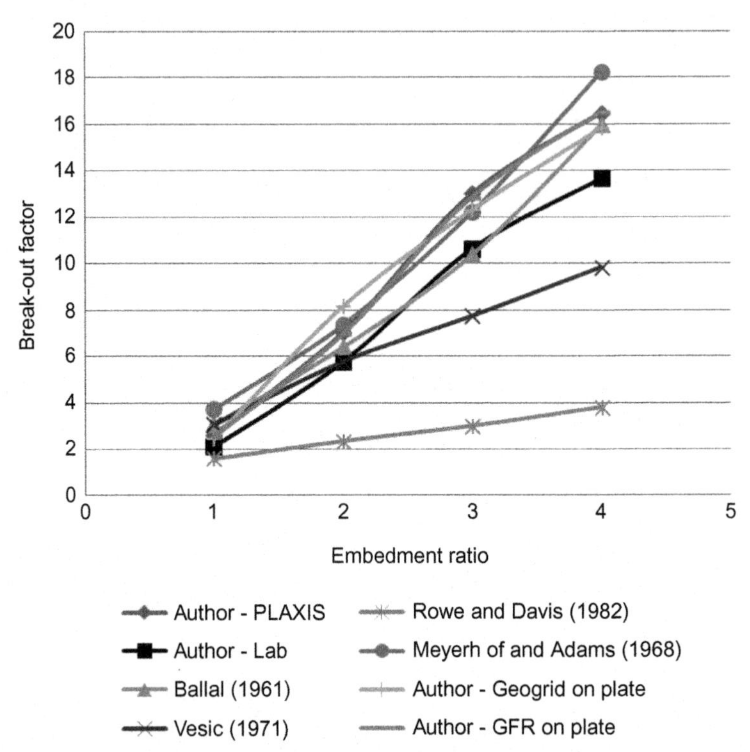

Figure 5.1 Comparison of breakout factor between experimental results and theoretical and numerical prediction for circular anchor plates in loose packing.

5.2.1 Symmetrical Anchor Plates in Loose Packing

Figs. 5.1–5.3 illustrate comparisons of theoretical breakout factor values and current results based on experimental and numerical analysis. The overall trend indicates that for the series of tests and models conducted, experimental and numerical values are in close agreement and are similar to the values of Balla (1961) for circular plates, Vesic (1971) for square plates, and Meyerhof and Adams (1968) for rectangular plates. For the symmetrical anchor plates shown in Figs. 5.3–5.6, the numerical results using PLAXIS values are seen to be much lower than the tested values of fiber-reinforced polymer breakout factors using laboratory tests, and are seen to be in agreement with values of Balla (1961) for square plates and Meyerhof and Adams (1968) for circular and rectangular plates.

The figures also show that the breakout factor increases significantly with an increase in the anchor embedment depth for grid fixed reinforcement (GFR) reinforced, reinforced, and non-reinforced soil. They also

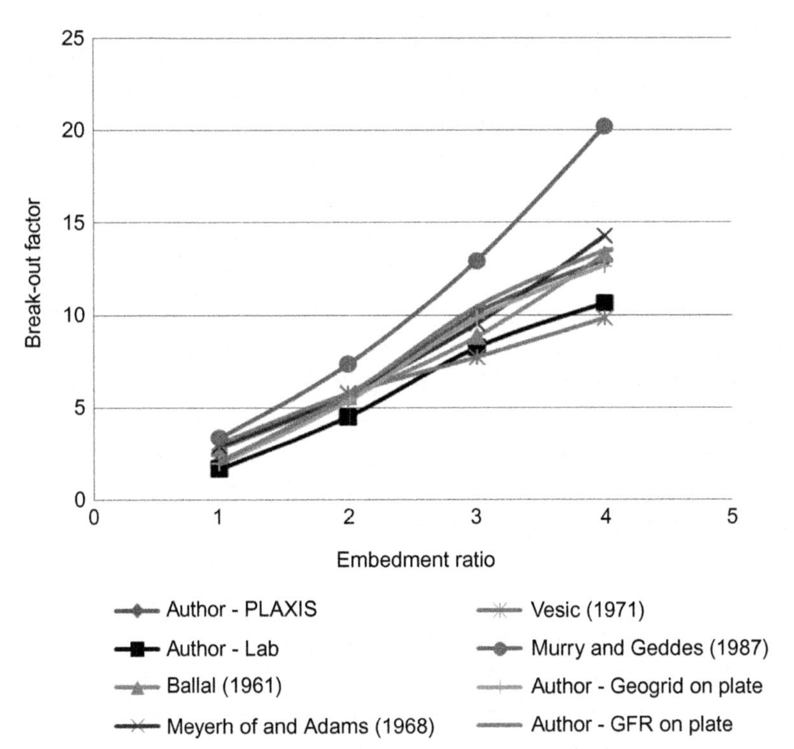

Figure 5.2 Comparison of breakout factor between experimental results and theoretical and numerical prediction for square anchor plates in loose packing.

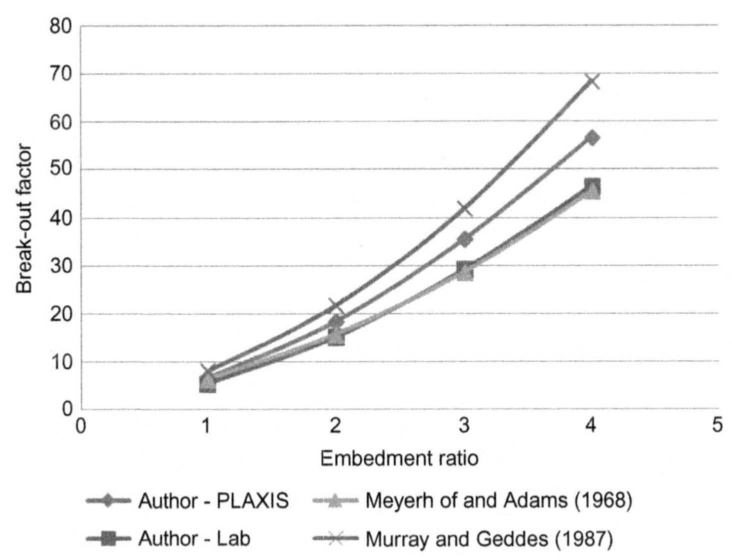

Figure 5.3 Comparison of breakout factor between experimental results and theoretical and numerical prediction for rectangular anchor plates in loose packing.

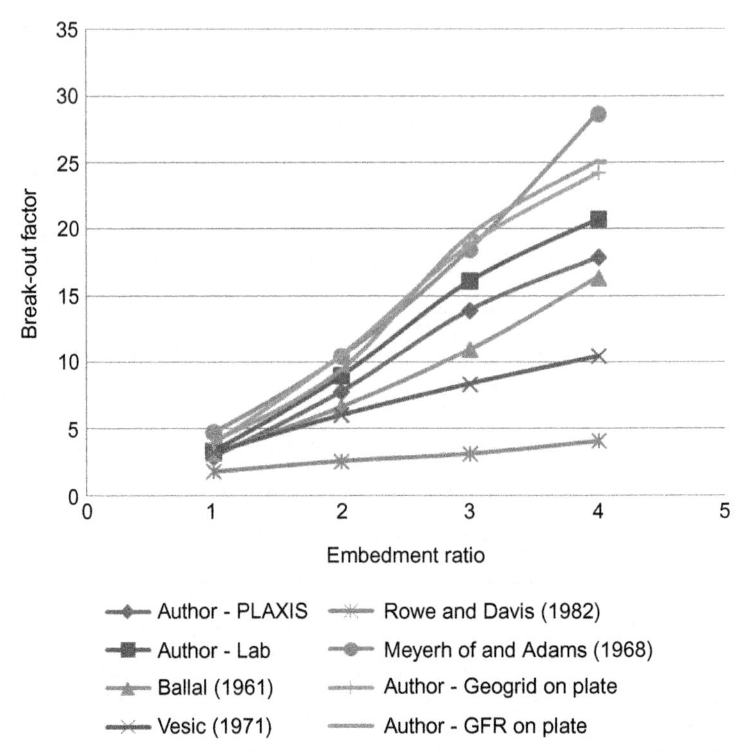

Figure 5.4 Comparison of breakout factor between experimental results and theoretical and numerical prediction for circular anchor plates in dense packing.

show that the inclusion of one geogrid layer over the symmetrical anchor plate results in a greater breakout factor than that produced when the same symmetrical anchor plate is embedded without reinforcement. It can also be noted that a GFR layer over the symmetrical anchor plate results in a greater breakout factor than the same conditions with a geogrid layer.

5.2.2 Symmetrical Anchor Plates in Dense Packing

Comparisons of experimental with theoretical or numerical predictions for symmetrical anchor plates in dense packing are shown in Figs. 5.4–5.6. The values of the breakout factor from the experimental laboratory testing are in close agreement and similar to the values of Balla (1961) for circular and square anchor plates and Meyerhof and Adams (1968) for strip anchor plates. For symmetrical anchor plates, shown in Figs. 5.3–5.6, numerical results using PLAXIS values are seen to be

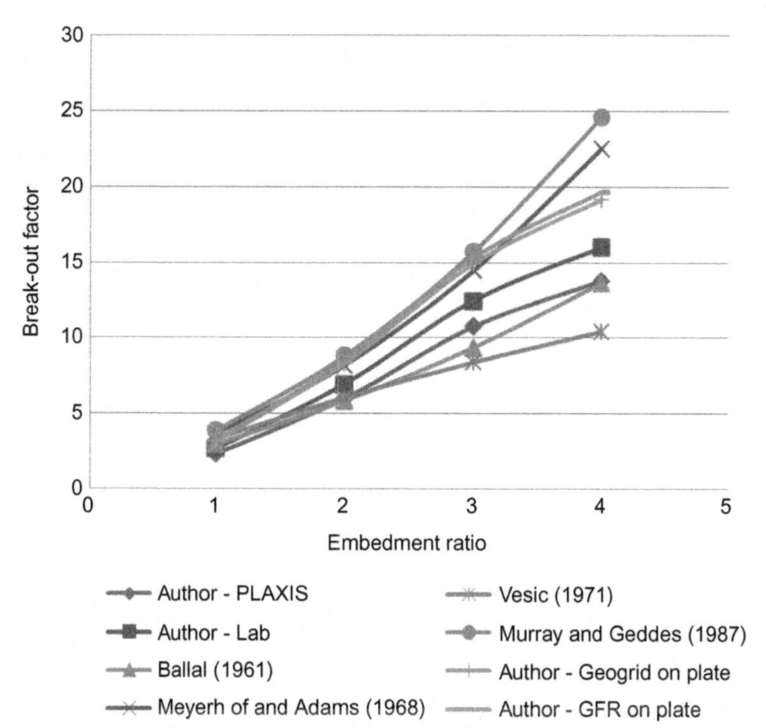

Figure 5.5 Comparison of breakout factor between experimental results and theoretical and numerical prediction for square anchor plates in dense packing.

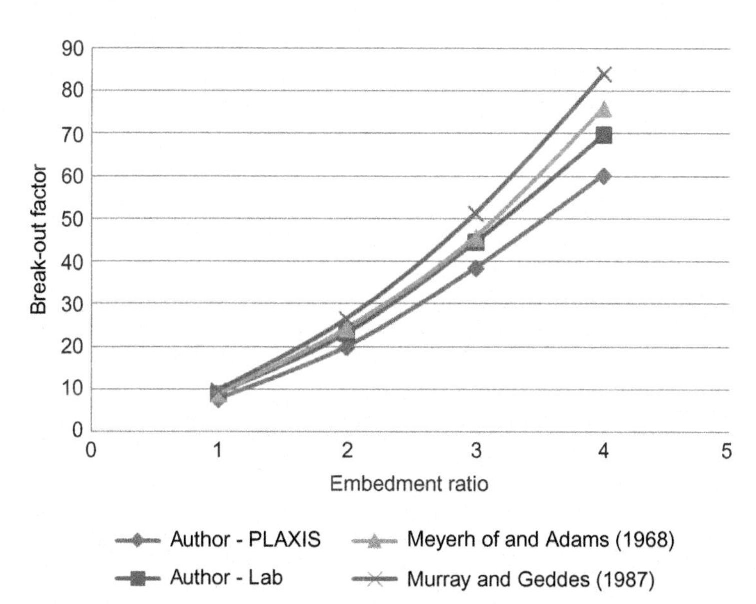

Figure 5.6 Comparison of breakout factor between experimental results and theoretical and numerical prediction for rectangular anchor plates in dense packing.

much lower than the values gathered using laboratory tests. These are similar to the values of Balla (1961) for circular and square anchor plates and Meyerhof and Adams (1968) for rectangular anchor plates.

Figs. 5.4 and 5.5 show that the breakout factor increases significantly with the anchor embedment depth for GFR, reinforced and not reinforced cases. Also, inclusion of one geogrid layer over the symmetrical anchor plate gives a greater breakout factor than that produced when the same symmetrical anchor plate is embedded without reinforcement. It is also noticed that a GFR layer over the symmetrical anchor plate gives a greater breakout factor than the same conditions with a geogrid layer.

5.3 CONCLUSION

No overall agreement was found between the author's results and existing theories, which include the effects of reinforcement materials and embedment ratios. This result was expected because different researchers employed different parameters in their researches, although close agreement for the breakout factor and embedment ratio with previous existing theories was found. The author's predictions of breakout factors for all ranges of embedment depth to diameter ratios (L/D) are in close agreement with Balla's estimation curve in circular and square plates and with Meyerhof's prediction.

REFERENCES

Balla, A. (1961). The resistance of breaking-out of mushroom foundations for pylons. *Proceedings of the Fifth International Conference on Soil Mechanics and Foundation Engineering, 1*, 569–576.

Dickin, E. A., & Laman, M. (2007). Uplift response of strip anchors in cohesionless soil. *Journal of Advanced Software Engineering, 1*(38), 618–625.

Frgic, L., & Marovic, P. (2003). Pullout capacity of spatial anchors. *Journal of Engineering Computations, 21*(6), 598–700.

Krishna, Y. S. R. (2000). *Numerical analysis of large size horizontal strip anchors*. PhD Thesis, Indian Institute of Science.

Kumar, J., & Bhoi, M. K. (2008). Interference of multiple strip footings on sand using small scale model tests. *Geotechnical and Geological Engineering, 26*(4), 469–477.

Kuzer, K. M., & Kumar, J. (2009). Vertical uplift capacity of two interfering horizontal anchors in sand using an upper bound limit analysis. *Computers and Geotechnics, 1*(36), 1084–1089.

Merifield, R., & Sloan, S. W. (2006). The ultimate pullout capacity of anchors in frictional soils. *Canadian Geotechnical Journal, 43*(8), 852–868.

Meyerhof, G. G., & Adams, J. I. (1968). The ultimate uplift capacity of foundations. *Canadian Geotechnical Journal, 5*(4), 225–244.
Murray, E. J., & Geddes, J. D. (1987). Uplift of anchor plates in sand. *Journal of Geotechnical Engineering, 113*(3), 202–215.
Rowe, R. K., & Davis, E. H. (1982). The behaviour of anchor plates in sand. *Geotechnique, 32*(1), 25–41.
Sarac, D. Z. (1989). Uplift capacity of shallow buried anchor slabs. *Proceedings of the 12th International Conference on Soil Mechanics and Foundation Engineering, 12*(2), 1213–1218.
Smith, C. C. (1998). Limit loads for an anchor/trapdoor embedded in an associated coulomb soil. *Int J Numer Anal Methods Geomech, 22*(11), 855–865.
Vesic, A. S. (1971). Breakout resistance of objects embedded in ocean bottom. *Journal of the Soil Mechanics and Foundations Division, 97*(9), 1183–1205.

CHAPTER 6

Conclusions and Recommendations

6.1 INTRODUCTION

This research provided new information on symmetrical anchor plates with soil reinforcement and grid fixed reinforcement (GFR). Although it is not dedicated to any specific practical conditions in engineering practice, it is useful to research the various factors that influence the symmetrical anchor plate's capacity when it is subjected to uplift forces. An analysis of the effects of various parameters has enabled a deeper understanding of the behavior of reinforced and grid fixed reinforced symmetrical anchor plates in sand when subjected to uplift loads. The failure shape for symmetrical anchor plates with embedment depth (L/D) of up to 4 is cylindrical despite variation in size, density, and reinforcement materials in reinforced and grid fixed reinforced anchor plates in sand when subjected to uplift loads.

In the selection of a symmetrical anchor plate's depth to achieve an optimum anchor plate uplift design, the size and depth are important parameters to be taken into consideration. It would therefore be more economical and rational to increase the uplift capacity of a symmetrical anchor plate by increasing its depth. An increase in the symmetrical anchor plate's depth increases the uplift capacity more significantly than compared to an increase in the symmetrical anchor plate's size by increasing the anchor plate contact area with the soil.

A rectangular anchor plate provides a higher uplift response compared with square or circular anchor plates. A deeply embedded strip-shaped symmetrical anchor plate would be substantially more resistant to uplift forces than a square- or circular-shaped symmetrical anchor plate. This increased resistance is due to the geometric progression in capacity with the increasing depth of the symmetrical anchor plate. This helps in the selection of the shape of a symmetrical anchor plate to achieve an economic anchor plate uplift design.

Soil Reinforcement for Anchor Plates and Uplift Response.
DOI: http://dx.doi.org/10.1016/B978-0-12-809558-4.00006-6
© 2017 Elsevier Inc.
All rights reserved.

It is also important to note that soil packing was found to be the most influential parameter in increasing uplift capacity. Adequate compaction of soils around the symmetrical anchor plates is an important factor, as indicated by the tests conducted on soils with relatively increased density.

The number and vertical spacing of the geogrid layers are important for the design of symmetrical anchor plates. In fact, the inclusion of one geogrid layer resting directly on top of the anchor plate approximately produces the same effect as that produced by including multiple geogrid layers. However, for a special design for the anchor plates using multiple geogrid layers, the optimum vertical spacing between the geogrid layers should be $0.5D$.

Increasing the frictional surface of GFR to soil will invariably cause geometric increases in the uplift capacity. It is concluded that, in terms of the symmetrical anchor plate's capacity, multiple layers of GFR are not more effective than reinforcing the soil itself with one layer. It would, therefore, be more economical and rational to increase the uplift capacity of symmetrical anchor plates by GFR because GFR decreases the geogrid area, which is very cost effective.

The outfitted tension trend in the reinforcement allows the geogrid layer to resist the formed horizontal shear stresses built up in the soil mass inside the loaded zone, and its movement beside the stable layers of loosely and densely packed soil leading to a broader and deeper failure zone. Based on this result, sand–geogrid interaction not only achieves an increase in the uplift force due to developed longer failure surface but also extends the contact zone between the soil and the laboratory box.

Empirical expression developed by the author gives slightly different values than those derived from experimental works. This change in values is due to the rounding up of the approximate values to derive the overall empirical expressions for symmetrical anchor plates and the coefficient of dimensional factor. The author's empirical expression results for symmetrical anchor plates in loose-packed soil for overall range uplift procedure are in agreement with the prediction given by Balla (1961). The theory proposed by Meyerhof and Adams (1968) with loose-packed soil for a range of L/D ratios gives a higher value than the author's prediction. In densely packed soil conditions, the author's empirical expression value shows an agreement with the prediction curve given by Balla (1961) for all ranges of L/D ratios, although Meyerhof's curve gives the highest values for all ranges of L/D ratios.

Test and numerical simulation results showed that using geogrid and GFR reinforcement has a significant effect on improving the uplift capacity of symmetrical anchor plates. It was found that inclusion of one layer of geogrid on the symmetrical anchor plate improved the uplift capacity by 20% as compared to the same symmetrical anchor plate embedded without reinforcement. However with the inclusion of GFR, the uplift response improved further to 30%. The results showed that the use of GFR to reinforce the cohesionless soil increased the uplift capacity of symmetrical anchor plates by a factor of 1.3- and 1.2-times compared with that for non-reinforced cohesionless soil and reinforced soil using geogrid, respectively.

From the author's point of view, it is convenient to predict the uplift capacity and breakout factor in sand for symmetrical anchor plates based on geogrid and GFR with the association of reinforcement parameters. The function of the reinforcement parameters developed by the author is more appropriate than those developed by others. This is due to the non-linear derivation of the author's analyses which is more relevant to the behavior of the soil properties. However to simplify the analyses, the author adopted the method of Meyerhof and Adams (1968) by transforming the non-linear derivation to a linear relationship. Thus an average line is developed among the non-linear data obtained. Regression analysis was used to obtain a best-fitting straight line. By employing the linear equation, the correlation of coefficient of dimensionless factor a solution is obtained for interpreting the soil reinforcement parameters in uplift response.

6.2 SHORTCOMINGS OF THE RESEARCH

The application of the proposed guidelines needs to be verified by prototype anchor plates to confirm or to provide an indication of its reliability. Due to scale effects the data may need modification since the data were obtained from model laboratory testing or numerical simulation and not actual site results.

6.3 RECOMMENDATIONS FOR FUTURE RESEARCH

- A similar research conducted for symmetrical anchor plates with cohesive soils.
- A similar research conducted for symmetrical anchor plates with other types of geosynthetics such as geotextiles, geonets, geojackets, geomembranes, and a wider range of soil parameters.
- Practical investigation on site to obtain data on uplift response for current research.

- A similar research conducted for symmetrical anchor plates with various sizes of GFR based on tie lengths and cubic dimensions.
- A similar research conducted on symmetrical anchor plates with various sizes of GFR based on tie lengths and cubic dimensions and other types of geosynthetics on multilayer soils.
- A similar research conducted on symmetrical anchor plates with various sizes of GFR based on tie lengths and cubic dimensions and other types of geosynthetics on cohesive soils.
- A similar research conducted on symmetrical anchor plates with various sizes of GFR based on tie lengths and cubic dimensions and other types of geosynthetics by numerical analysis based on numerical methods, such as the discrete element method, the finite difference method, etc.

REFERENCES

Balla, A. (1961). The resistance of breaking-out of mushroom foundations for pylons. *Proceedings of the Fifth International Conference on Soil Mechanics and Foundation Engineering, 1*, 569–576.

Meyerhof, G. G., & Adams, J. I. (1968). The ultimate uplift capacity of foundations. *Canadian Geotechnical Journal, 5*(4), 225–244.

APPENDIX A

Grid Fixed Reinforcement Design in ANSYS

UNITS

Table A.1

Unit system	Metric (m, kg, N, °C, s, V, A)
Angle	Degrees
Rotational velocity	rad/s

MODEL

Geometry

Table A.2 Model > geometry

Object name	***Geometry***
State	Fully defined
Definition	
Source	E.\ANSYS RUNS\Workspace \Sina\Sina-Strip.agdb
Type	Design modeler
Length unit	Millimeters
Element control	Program controlled
Display style	Part color
Bounding box	
Length X	4.0063e − 002 m
Length Y	1.e − 002 m
Length Z	8.1666e − 002 m
Properties	
Volume	1.0598e − 006 m^2
Mass	1.0068e − 003 kg

(*Continued*)

Statistics	
Bodies	1
Active bodies	1
Nodes	2.499
Elements	1036
Preferences	
Import solid bodies	Yes
Import surface bodies	Yes
Import line bodies	Yes
Parameter processing	Yes
Personal parameter key	DS
CAD attribute transfer	No
Named selection processing	No
Material properties transfer	No
CAD associativity	Yes
Import coordinate systems	No
Reader save part file	No
Import using instances	Yes
Do smart update	No
Attach file via temp file	No
Analysis type	3D
Mixed import resolution	None
Enclosure and symmetry processing	Yes

Table A.3 Model > geometry > parts

Object name	***Part1***
State	Meshed
Graphics properties	
Visible	Yes
Transparency	1
Definition	
Suppressed	No
Material	Polyethylene
Stiffness behavior	Flexible
Nonlinear material effects	Yes
Bounding box	
Length X	4.0063e − 002 m
Length Y	1.e − 002 m
Length Z	8.1666e − 002 m

(*Continued*)

Table A.3 (Continued)

Object name	***Part1***
Properties	
Volume	1.0598e − 006 m^2
Mass	1.0068e − 003 kg
Centroid X	−3.9236e − 003 m
Centroid Y	5.e − 003 m
Centroid Z	1.7332e − 003 m
Moment of inertia Ip1	3.1544e − 007 kg/m^2
Moment of inertia Ip2	3.2272e − 007 kg/m^2
Moment of inertia Ip3	2.1421e − 008 kg/m^2
Statistics	
Nodes	2499
Elements	1036

Mesh

Table A.4 Model > mesh

Object name	***Mesh***
State	Solved
Defaults	
Physics preference	Mechanical
Relevance	0
Advanced	
Relevance center	Coarse
Element size	Default
Shape checking	Standard mechanical
Solid element midside nodes	Program controlled
Straight-sided elements	No
Initial size seed	Active assembly
Smoothing	Low
Transition	Fast
Statistics	
Nodes	2499
Elements	1036

STATIC STRUCTURAL

Table A.5 Model > analysis

Object name	***Static structural***
State	Fully defined
Definition	
Physics type	Structural
Analysis type	Static structural
Options	
Reference temperature	22°C

Table A.6 Model > static structural > analysis settings

Object name	***Analysis settings***
State	Fully defined
Step controls	
Number of steps	1
Current step number	1
Step end time	1 s
Auto time stepping	Program controlled
Solver controls	
Solver type	Program controlled
Weak springs	Program controlled
Large detection	Off
Inertia relief	Off
Nonlinear controls	
Force convergence	Program controlled
Moment convergence	Program controlled
Displacement convergence	Program controlled
Rotation convergence	Program controlled
Line search	Program controlled
Output controls	
Calculate stress	Yes
Calculate strain	Yes
Calculate results at	All time points

(*Continued*)

Table A.6 (Continued)

Object name	***Analysis settings***
Analysis data management	
Solver files directory	E1ANSYS RUNS\Workspace\Sina\Sina-Strip Simulation Files\Static Structural\
Future analysis	None
Save ANSYS db	No
Delete unneeded files	Yes
Nonlinear solution	No

Table A.7 Model > static structural > loads

Object name	***Force***	***Fixed support***
State	Fully defined	
Scope		
Sooping method	Geometry selection	
Geometry	1 face	
Definition		
Define by	Components	
Type	Force	Fixed support
X Component	400 N (ramped)	
Y Component	0 N (ramped)	
Z Component	0 N (ramped)	
Suppressed	No	

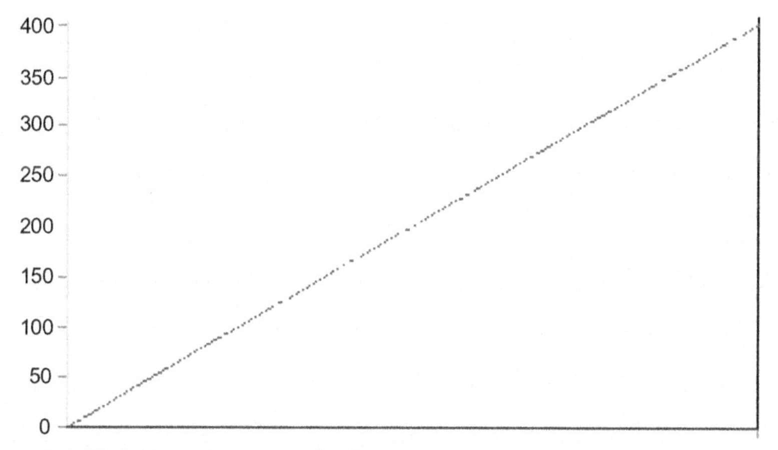

Figure A.1 Model > static structural > force.

Solution

Table A.8 Model > static structural > solution

Object name	***Solution***
State	Solved
Adaptive mesh refinement	
Max refinement loops	1
Refinement depth	2

Table A.9 Model > static structural > solution > solution information

Object name	***Solution information***
State	Solved
Solution information	
Solution output	Solver output
Newton–Raphson residuals	0
Update interval	2.5 s
Display points	All

Table A.10 Model > static structural > solution > results

Object name	***Total deformation***	***Directional deformation***	***Equivalent stress***
State	Solved		
Scope			
Geometry	All bodies		
Definition			
Type	Total deformation	Directional deformation	Equivalent (von-Mises) stress
Display time	End time		
Orientation		X-axis	
Results			
Minimum	0 m	−3.1716 m	19.713 Pa
Maximum	1450.3 m	1324.9 m	3.1284e + 011 Pa
Information			
Time	1 s		
Load step	1		
Substep	1		
Iteration number	1		

APPENDIX B

Data and Results

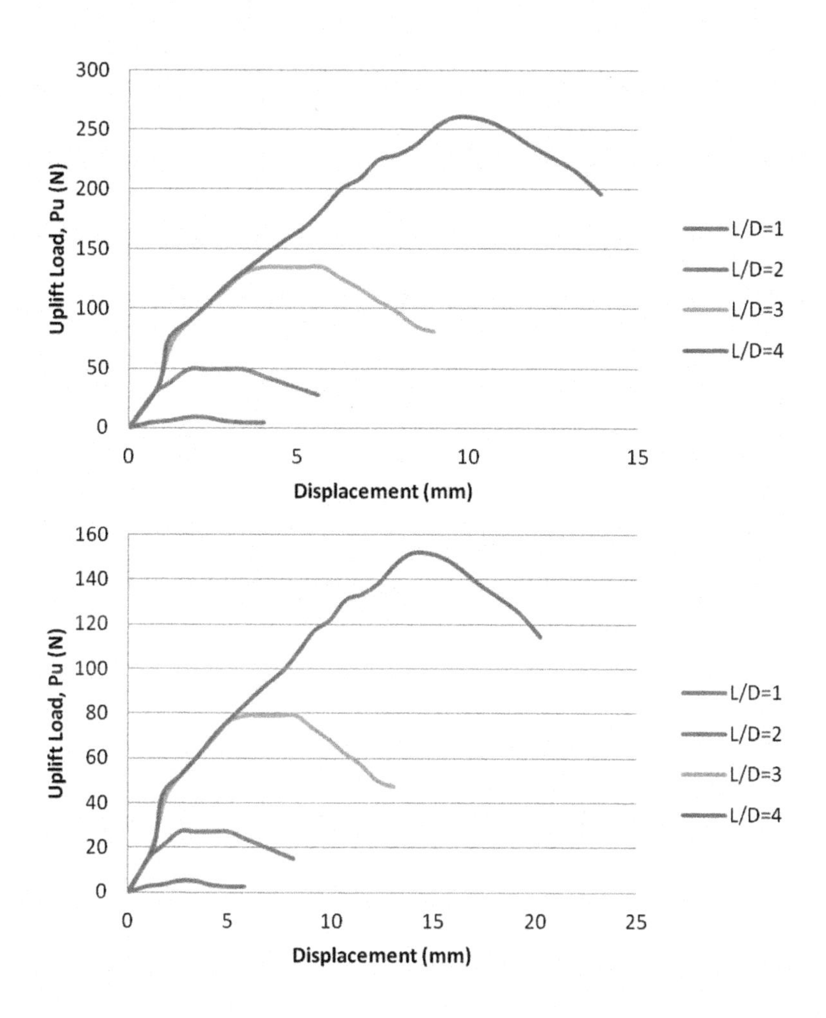

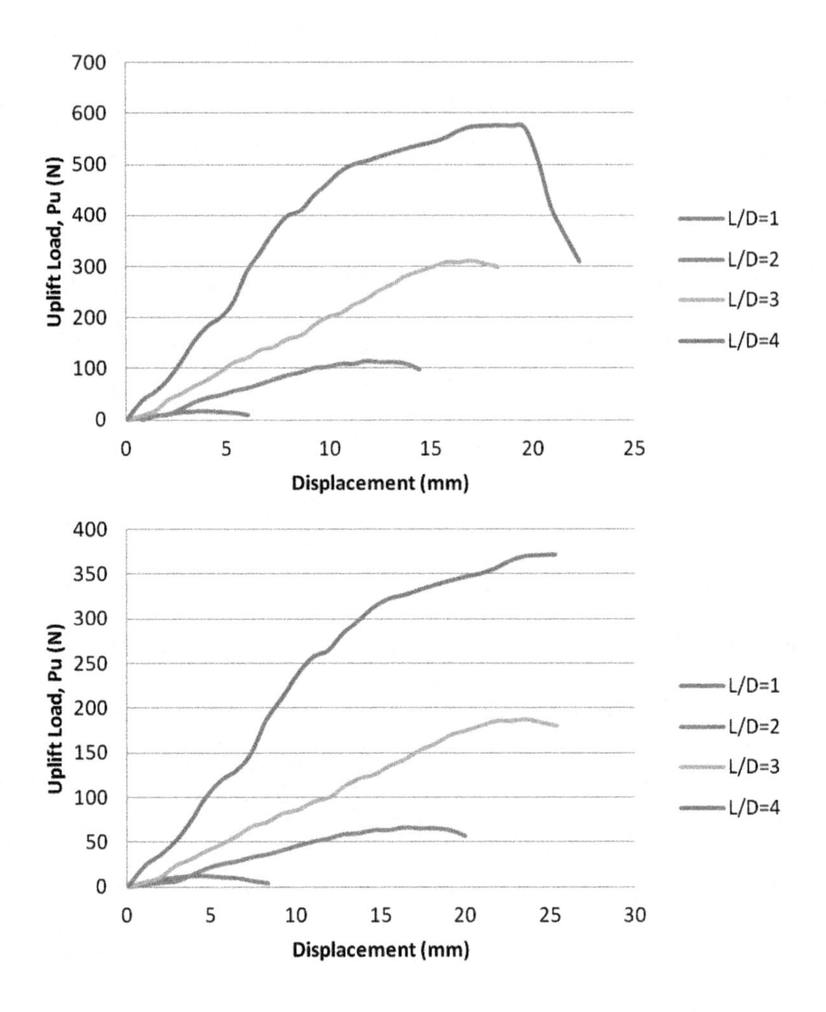
700
600
500
400
300
200
100
0
Uplift Load, Pu (N)
0
5
10
15
20
25
Displacement (mm)
L/D=1
L/D=2
L/D=3
L/D=4
400
350
300
250
200
150
100
50
0
Uplift Load, Pu (N)
0
5
10
15
20
25
30
Displacement (mm)
L/D=1
L/D=2
L/D=3
L/D=4

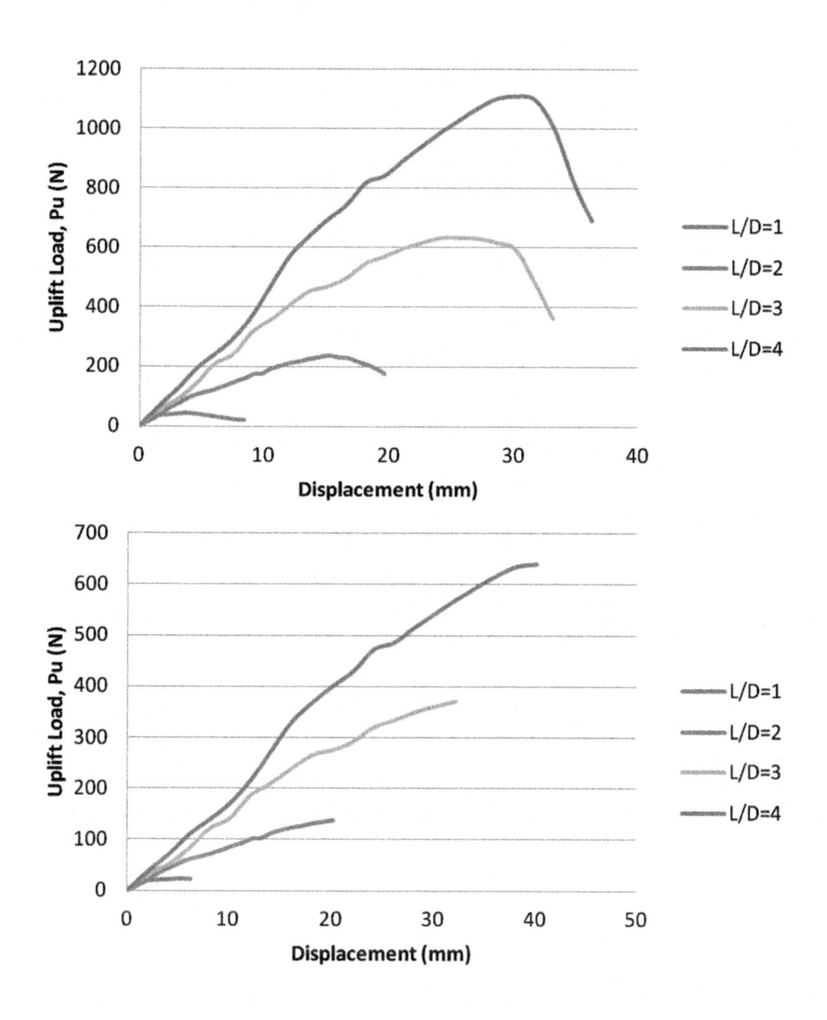
1200
1000
800
600
400
200
0
Uplift Load, Pu (N)
0
10
20
30
40
Displacement (mm)
L/D=1
L/D=2
L/D=3
L/D=4
700
600
500
400
300
200
100
0
Uplift Load, Pu (N)
0
10
20
30
40
50
Displacement (mm)
L/D=1
L/D=2
L/D=3
L/D=4

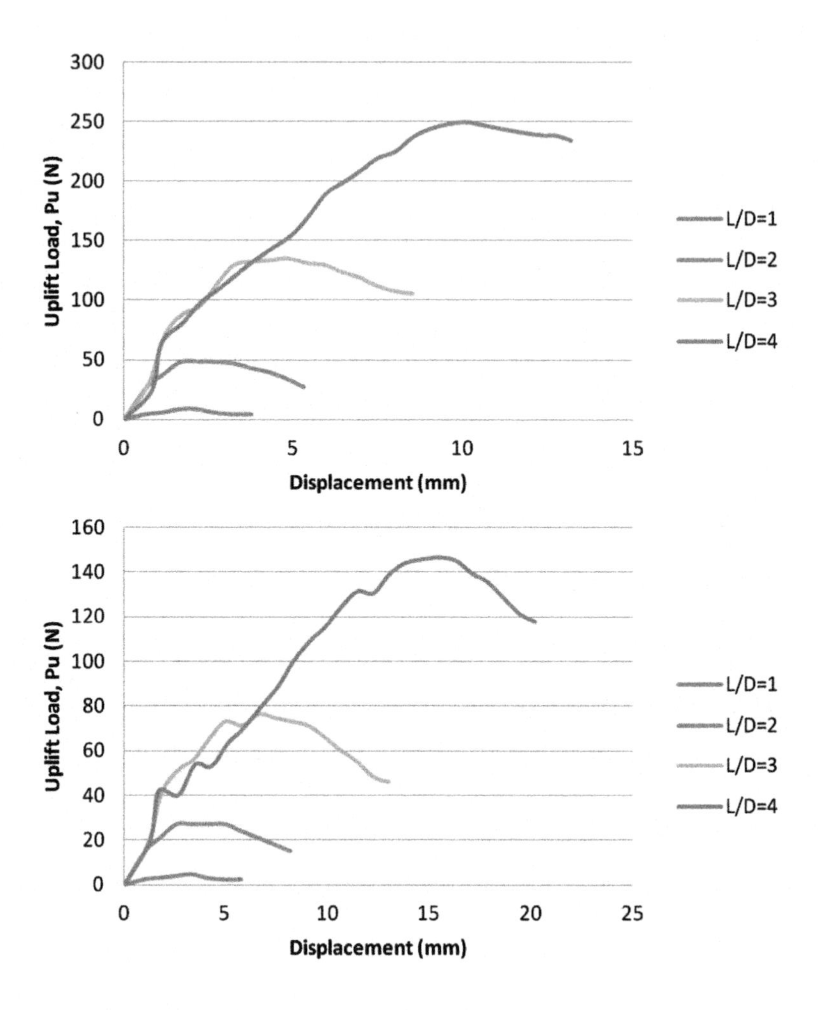
300
250
200
150
100
50
0
Uplift Load, Pu (N)
0
5
10
15
Displacement (mm)
L/D=1
L/D=2
L/D=3
L/D=4
160
140
120
100
80
60
40
20
0
Uplift Load, Pu (N)
0
5
10
15
20
25
Displacement (mm)
L/D=1
L/D=2
L/D=3
L/D=4

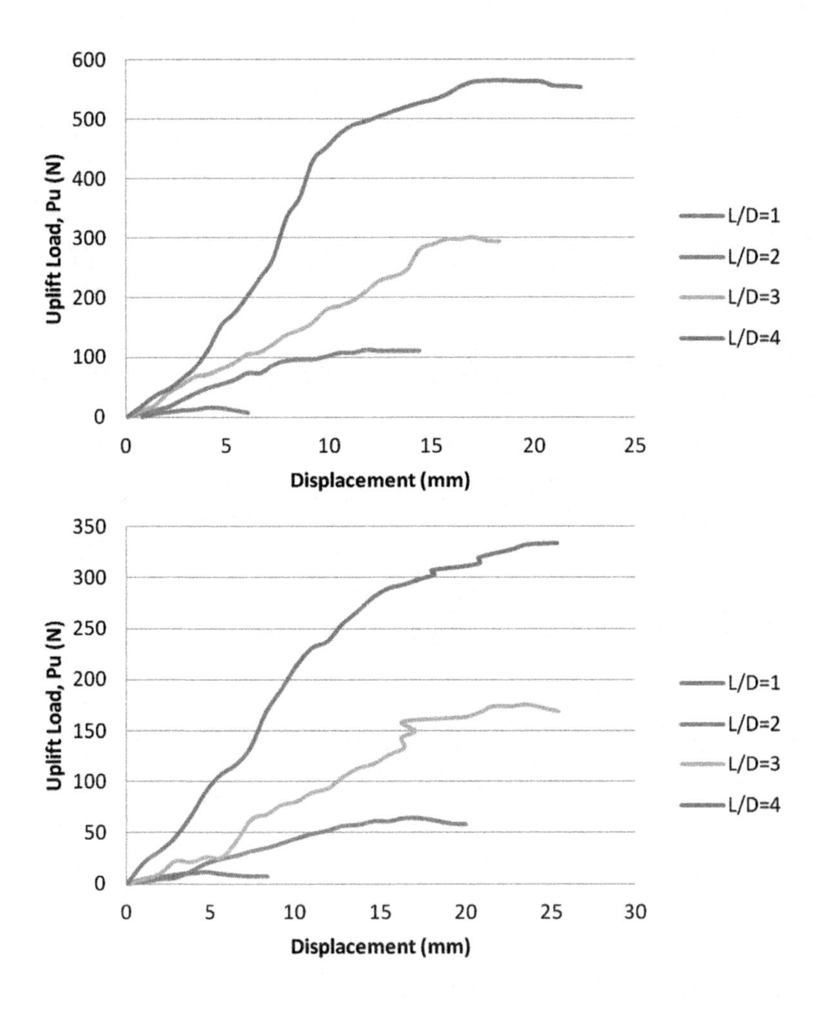
600
500
400
300
200
100
0
0
5
10
15
20
25
Uplift Load, Pu (N)
Displacement (mm)
L/D=1
L/D=2
L/D=3
L/D=4
350
300
250
200
150
100
50
0
0
5
10
15
20
25
30
Uplift Load, Pu (N)
Displacement (mm)
L/D=1
L/D=2
L/D=3
L/D=4

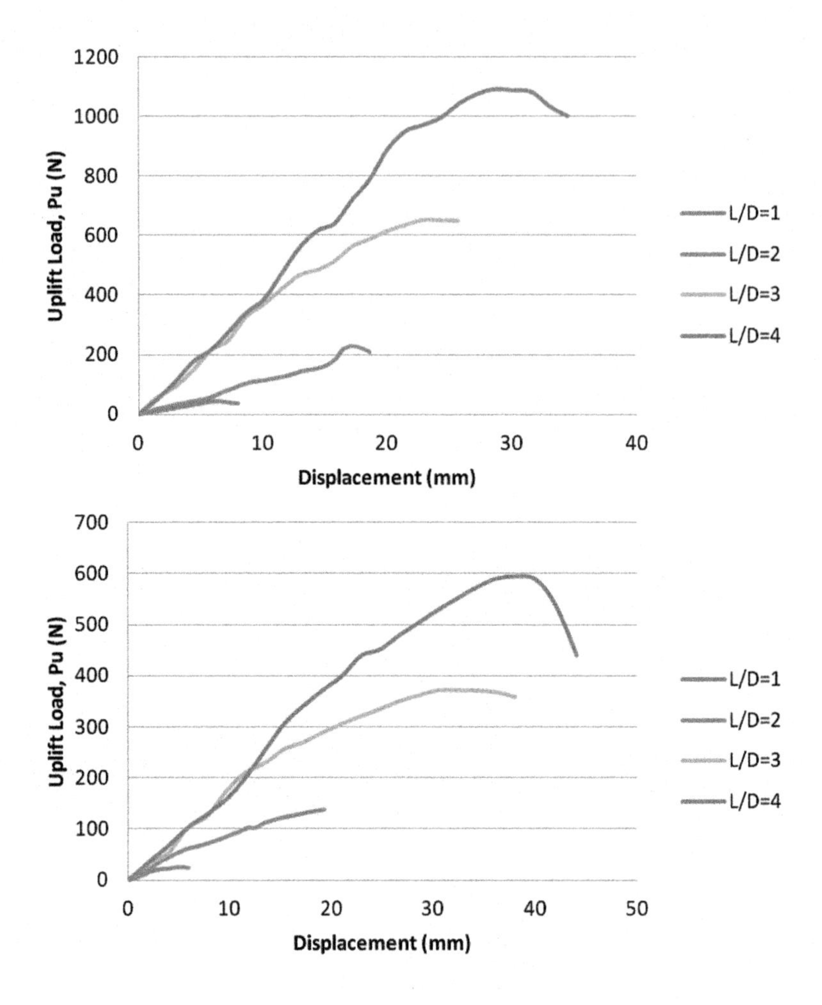
1200
1000
800
600
400
200
0
Uplift Load, Pu (N)
0
10
20
30
40
Displacement (mm)
L/D=1
L/D=2
L/D=3
L/D=4
700
600
500
400
300
200
100
0
Uplift Load, Pu (N)
0
10
20
30
40
50
Displacement (mm)
L/D=1
L/D=2
L/D=3
L/D=4

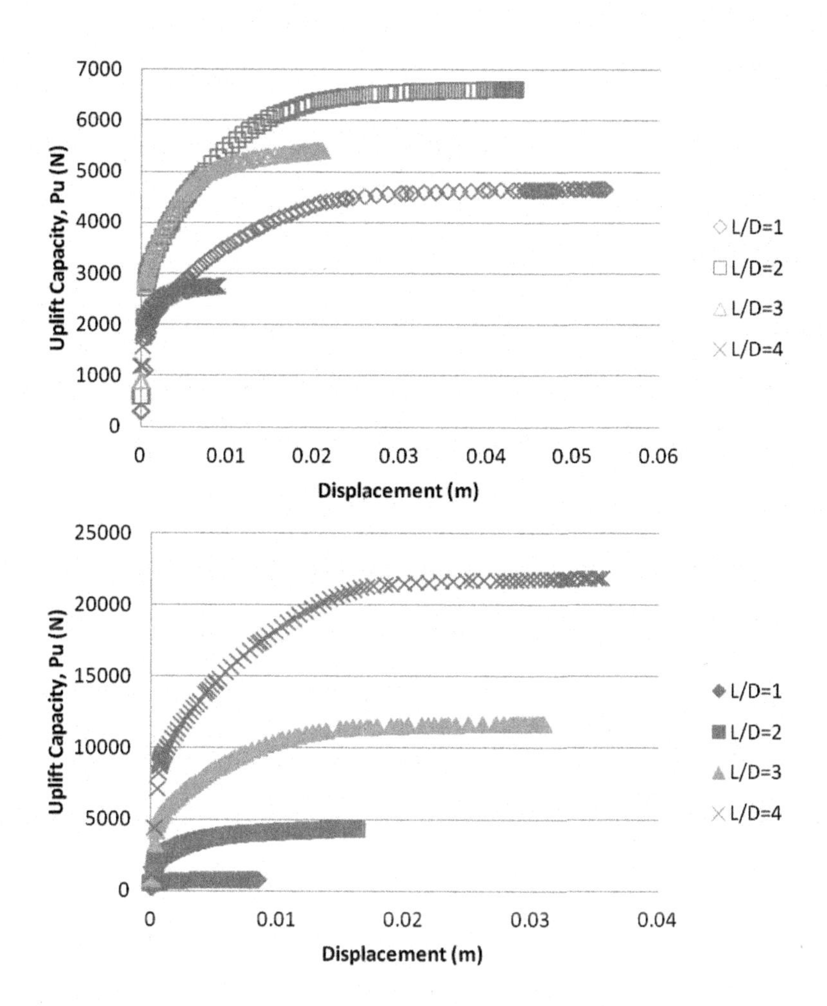
7000
6000
5000
4000
3000
2000
1000
0
Uplift Capacity, Pu (N)
0
0.01
0.02
0.03
0.04
0.05
0.06
Displacement (m)
L/D=1
L/D=2
L/D=3
L/D=4
25000
20000
15000
10000
5000
0
Uplift Capacity, Pu (N)
0
0.01
0.02
0.03
0.04
Displacement (m)
L/D=1
L/D=2
L/D=3
L/D=4

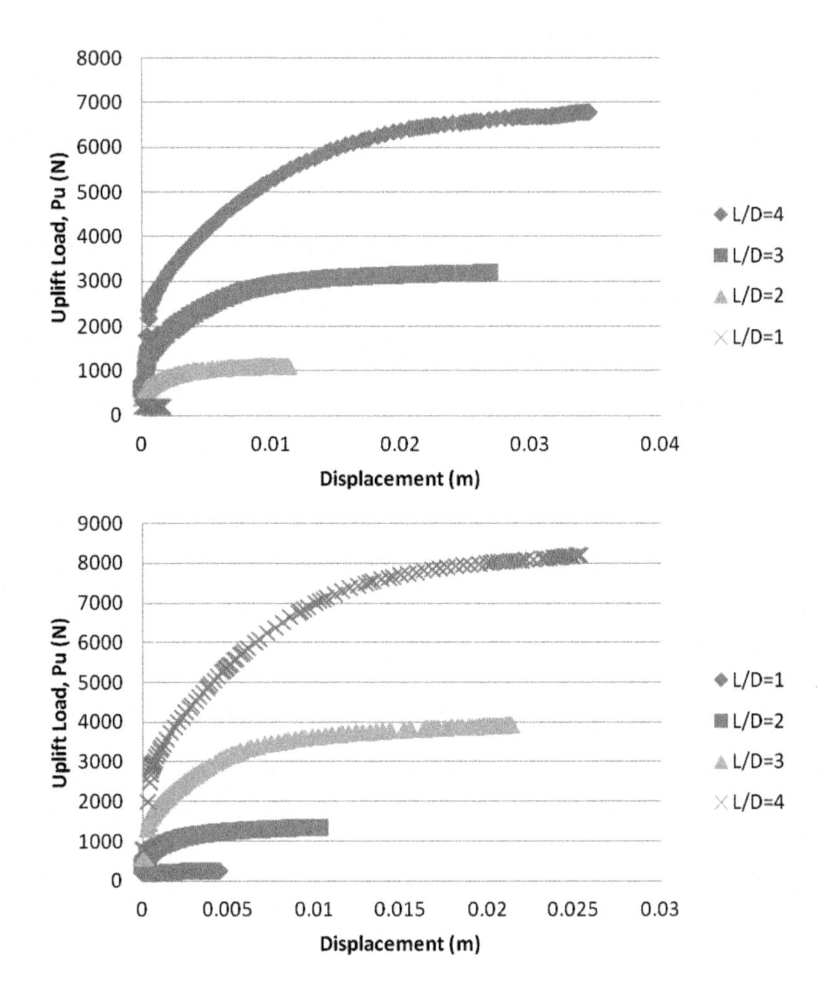
8000
7000
6000
5000
4000
3000
2000
1000
0
Uplift Load, Pu (N)
0
0.01
0.02
0.03
0.04
Displacement (m)
L/D=4
L/D=3
L/D=2
L/D=1
9000
8000
7000
6000
5000
4000
3000
2000
1000
0
Uplift Load, Pu (N)
0
0.005
0.01
0.015
0.02
0.025
0.03
Displacement (m)
L/D=1
L/D=2
L/D=3
L/D=4

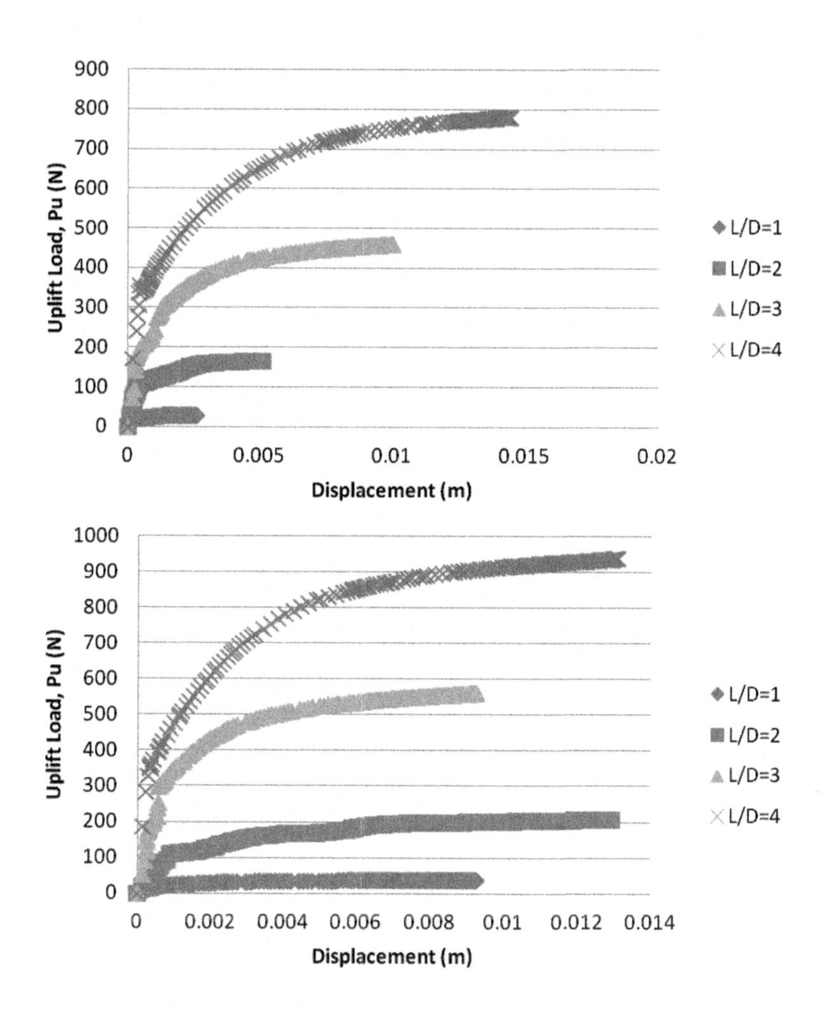
900
800
700
600
500
400
300
200
100
0
Uplift Load, Pu (N)
0
0.005
0.01
0.015
0.02
Displacement (m)
L/D=1
L/D=2
L/D=3
L/D=4
1000
900
800
700
600
500
400
300
200
100
0
Uplift Load, Pu (N)
0
0.002
0.004
0.006
0.008
0.01
0.012
0.014
Displacement (m)
L/D=1
L/D=2
L/D=3
L/D=4

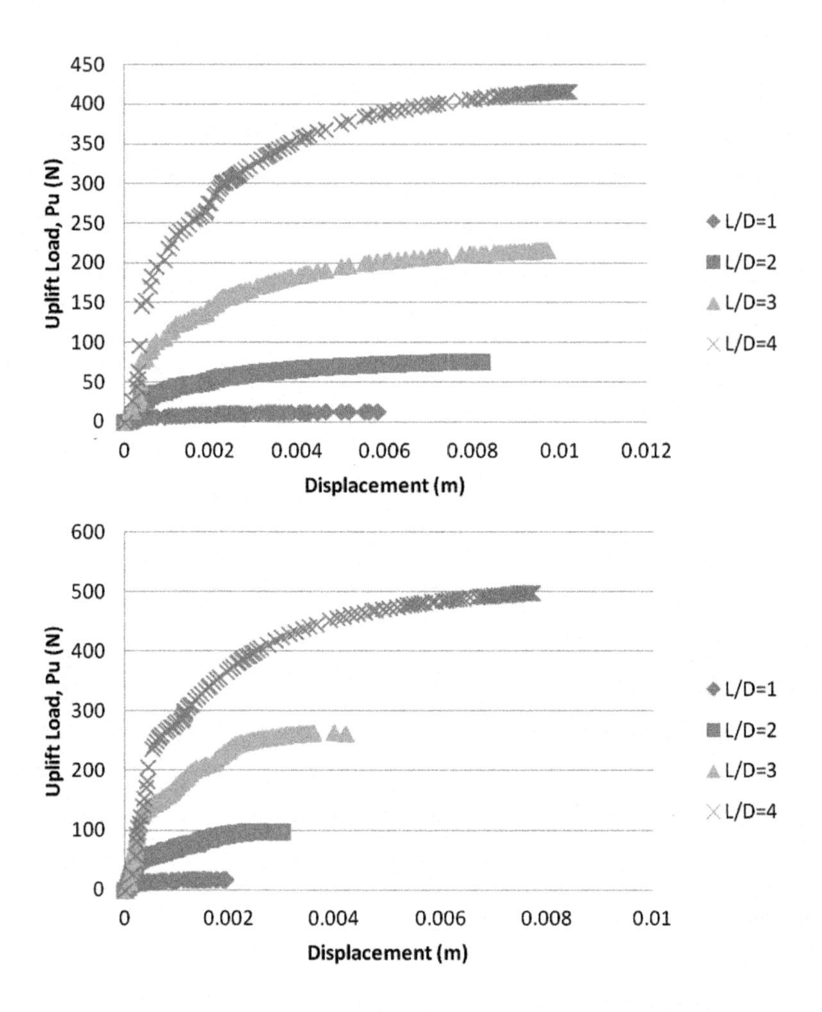
450
400
350
300
250
200
150
100
50
0
0
0.002
0.004
0.006
0.008
0.01
0.012
Uplift Load, Pu (N)
Displacement (m)
L/D=1
L/D=2
L/D=3
L/D=4
600
500
400
300
200
100
0
0
0.002
0.004
0.006
0.008
0.01
Uplift Load, Pu (N)
Displacement (m)
L/D=1
L/D=2
L/D=3
L/D=4

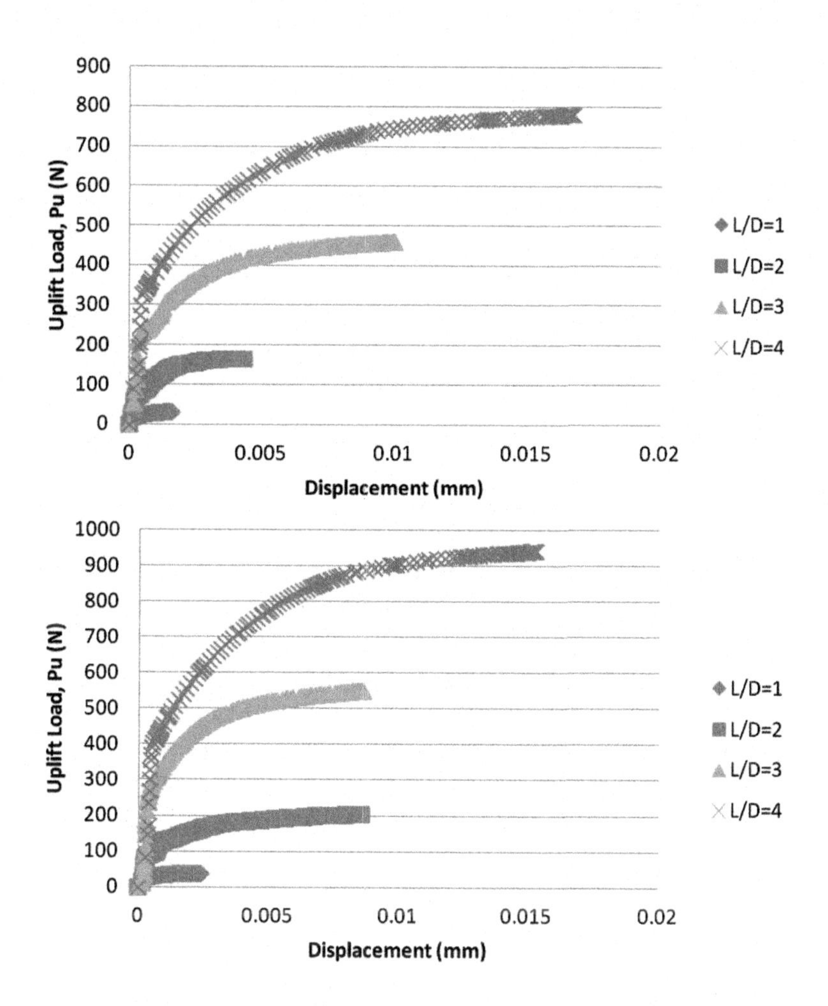
900
800
700
600
500
400
300
200
100
0
Uplift Load, Pu (N)
0
0.005
0.01
0.015
0.02
Displacement (mm)
L/D=1
L/D=2
L/D=3
L/D=4
1000
900
800
700
600
500
400
300
200
100
0
Uplift Load, Pu (N)
0
0.005
0.01
0.015
0.02
Displacement (mm)
L/D=1
L/D=2
L/D=3
L/D=4

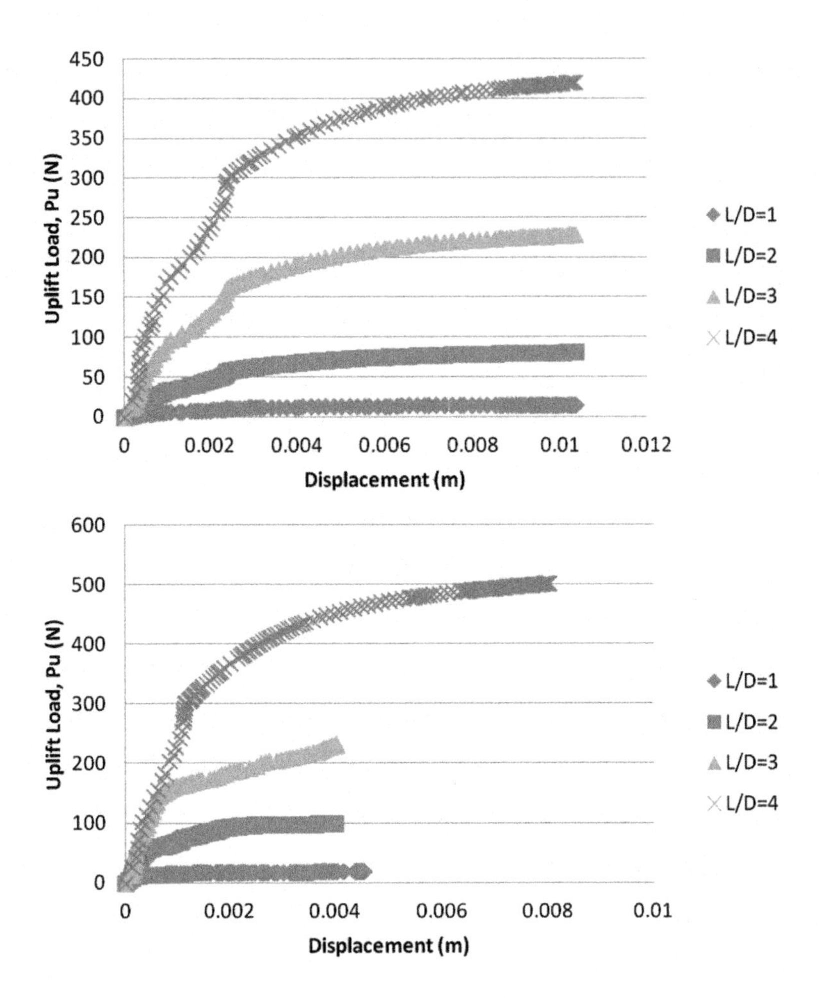
450
400
350
300
250
200
150
100
50
0
Uplift Load, Pu (N)
0
0.002
0.004
0.006
0.008
0.01
0.012
Displacement (m)
L/D=1
L/D=2
L/D=3
L/D=4
600
500
400
300
200
100
0
Uplift Load, Pu (N)
0
0.002
0.004
0.006
0.008
0.01
Displacement (m)
L/D=1
L/D=2
L/D=3
L/D=4

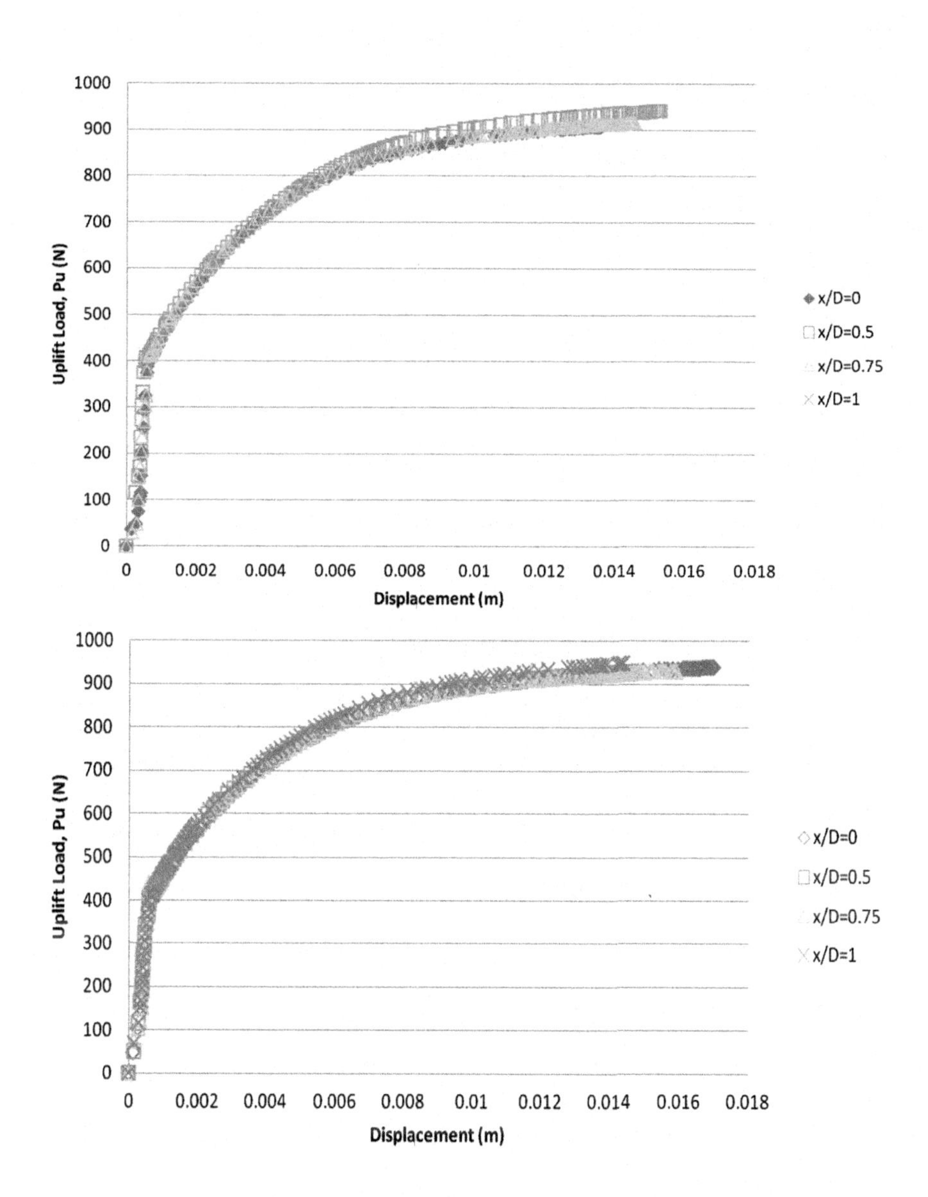

1000
900
800
700
600
500
400
300
200
100
0
Uplift Load, Pu (N)
0
0.002
0.004
0.006
0.008
0.01
0.012
0.014
0.016
0.018
Displacement (m)
x/D=0
x/D=0.5
x/D=0.75
x/D=1
1000
900
800
700
600
500
400
300
200
100
0
Uplift Load, Pu (N)
0
0.002
0.004
0.006
0.008
0.01
0.012
0.014
0.016
0.018
Displacement (m)
x/D=0
x/D=0.5
x/D=0.75
x/D=1

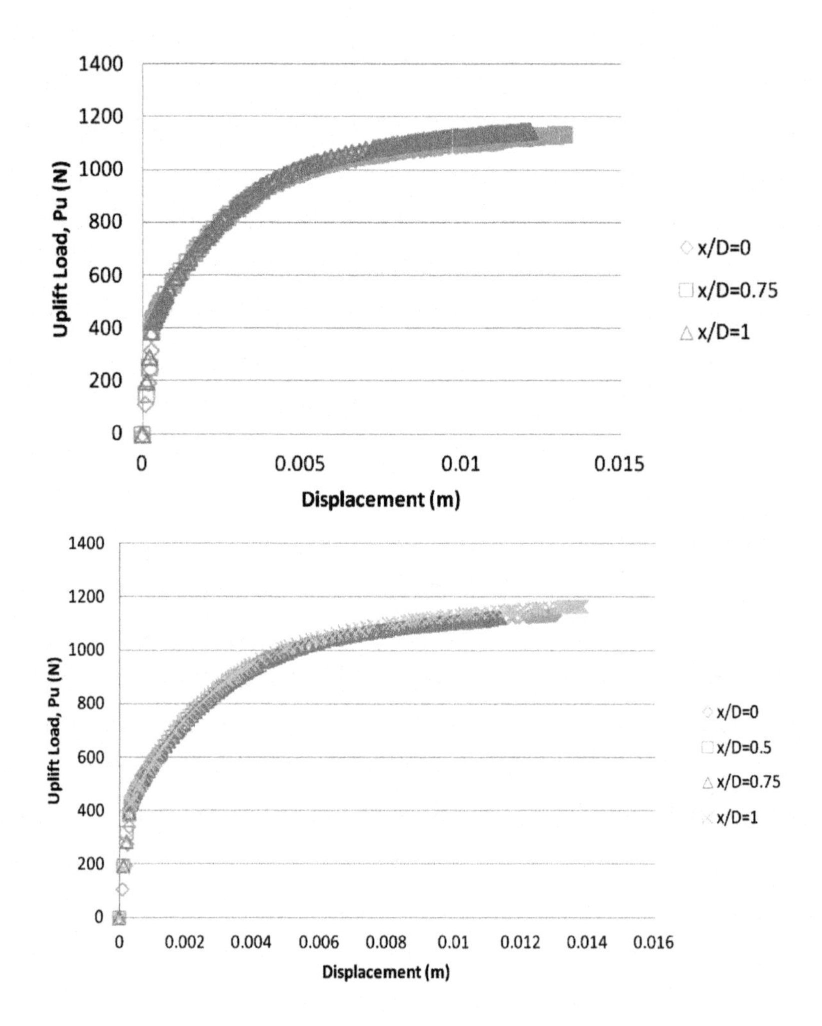
1400
1200
1000
800
600
400
200
0
Uplift Load, Pu (N)
0
0.005
0.01
0.015
Displacement (m)
x/D=0
x/D=0.75
x/D=1
1400
1200
1000
800
600
400
200
0
Uplift Load, Pu (N)
0
0.002
0.004
0.006
0.008
0.01
0.012
0.014
0.016
Displacement (m)
x/D=0
x/D=0.5
x/D=0.75
x/D=1

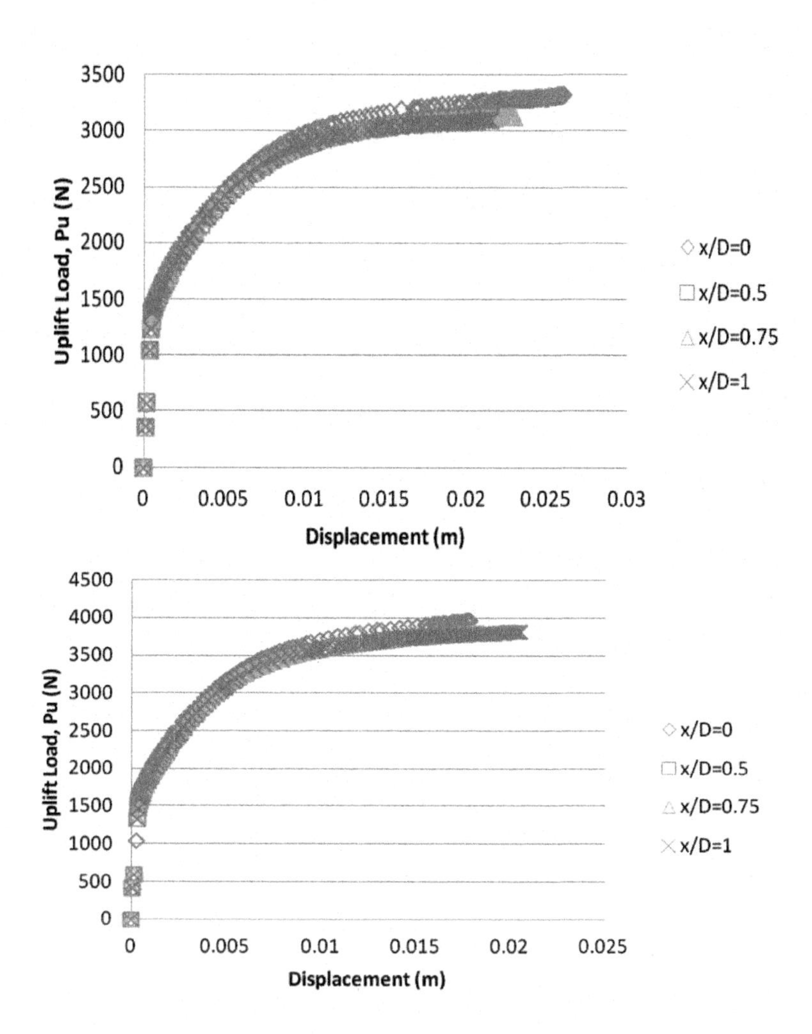
3500
3000
2500
2000
1500
1000
500
0
Uplift Load, Pu (N)
0 0.005 0.01 0.015 0.02 0.025 0.03
Displacement (m)
x/D=0
x/D=0.5
x/D=0.75
x/D=1
4500
4000
3500
3000
2500
2000
1500
1000
500
0
Uplift Load, Pu (N)
0 0.005 0.01 0.015 0.02 0.025
Displacement (m)
x/D=0
x/D=0.5
x/D=0.75
x/D=1

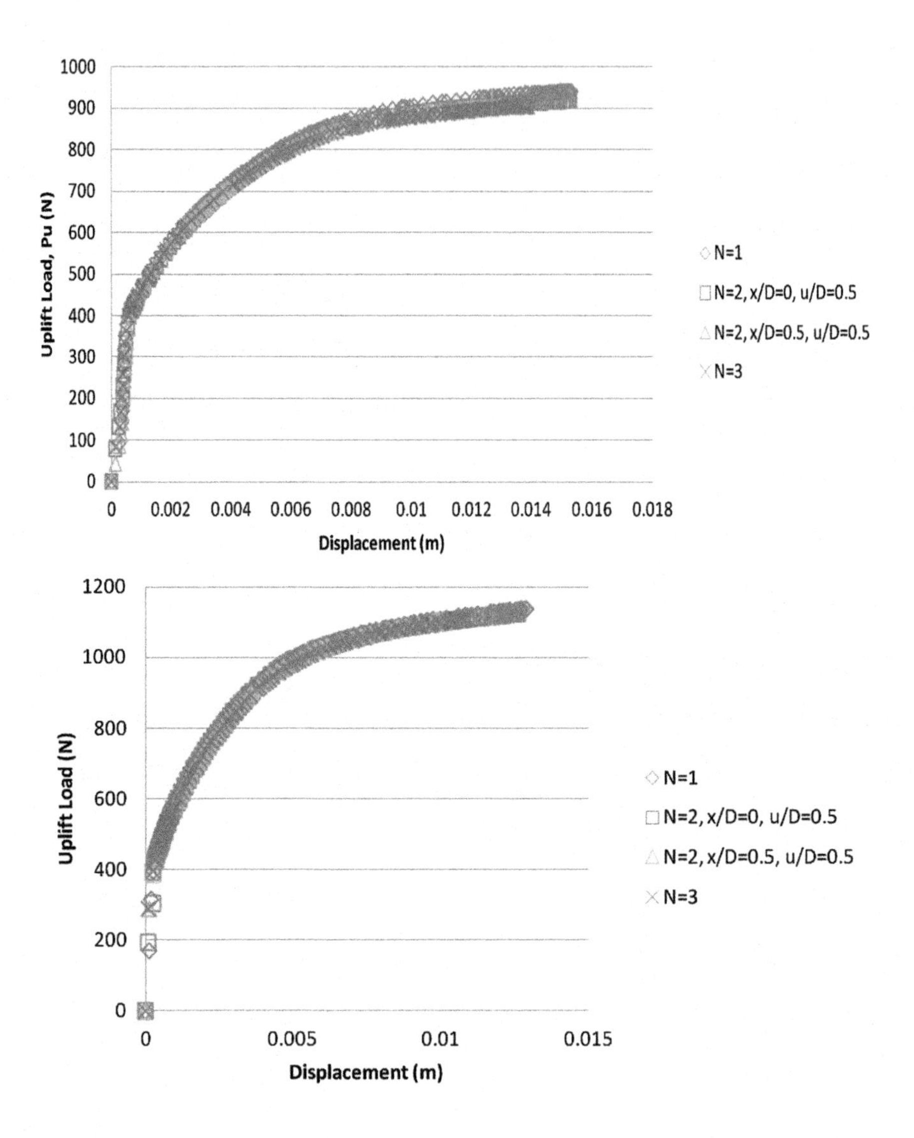

1000
900
800
700
600
500
400
300
200
100
0
Uplift Load, Pu (N)
0
0.002
0.004
0.006
0.008
0.01
0.012
0.014
0.016
0.018
Displacement (m)
N=1
N=2, x/D=0, u/D=0.5
N=2, x/D=0.5, u/D=0.5
N=3
1200
1000
800
600
400
200
0
Uplift Load (N)
0
0.005
0.01
0.015
Displacement (m)
N=1
N=2, x/D=0, u/D=0.5
N=2, x/D=0.5, u/D=0.5
N=3

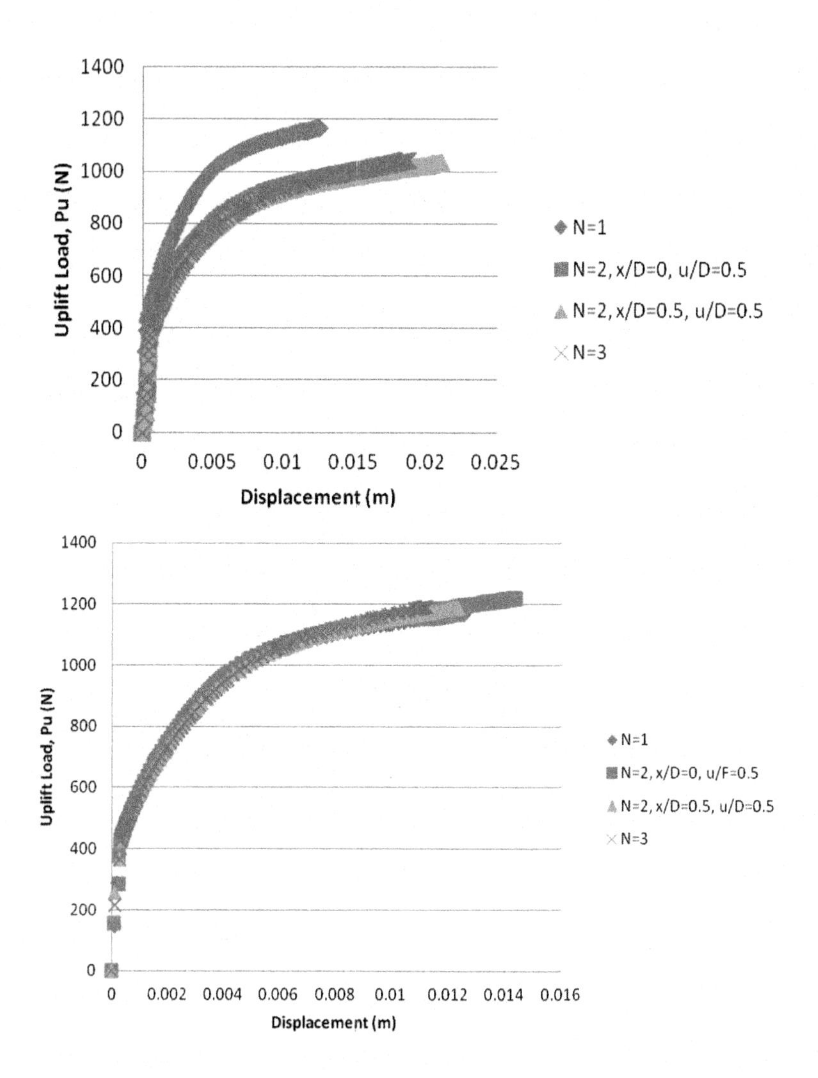
1400
1200
1000
800
600
400
200
0
Uplift Load, Pu (N)
0 0.005 0.01 0.015 0.02 0.025
Displacement (m)
N=1
N=2, x/D=0, u/D=0.5
N=2, x/D=0.5, u/D=0.5
N=3
1400
1200
1000
800
600
400
200
0
Uplift Load, Pu (N)
0 0.002 0.004 0.006 0.008 0.01 0.012 0.014 0.016
Displacement (m)
N=1
N=2, x/D=0, u/F=0.5
N=2, x/D=0.5, u/D=0.5
N=3

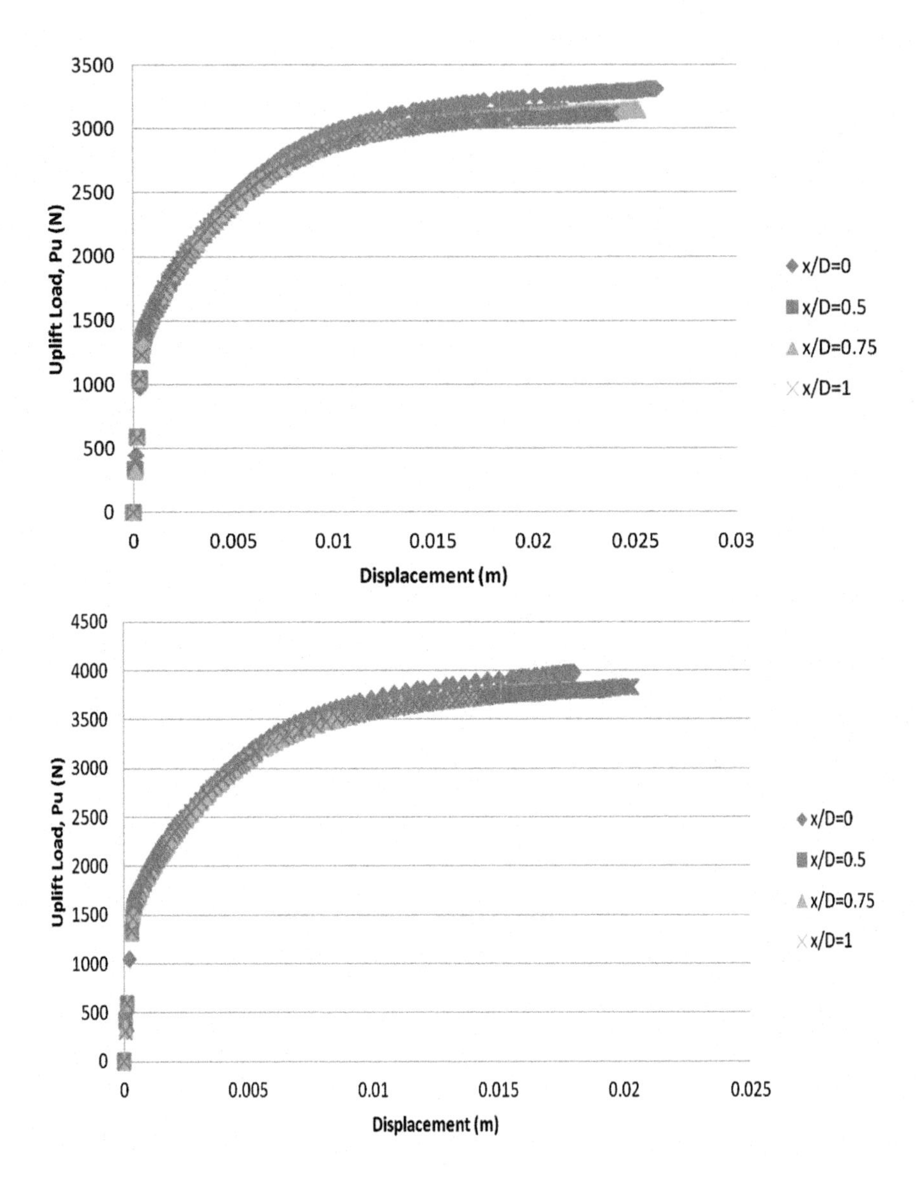

3500
3000
2500
2000
1500
1000
500
0
Uplift Load, Pu (N)
0
0.005
0.01
0.015
0.02
0.025
0.03
Displacement (m)
x/D=0
x/D=0.5
x/D=0.75
x/D=1
4500
4000
3500
3000
2500
2000
1500
1000
500
0
Uplift Load, Pu (N)
0
0.005
0.01
0.015
0.02
0.025
Displacement (m)
x/D=0
x/D=0.5
x/D=0.75
x/D=1

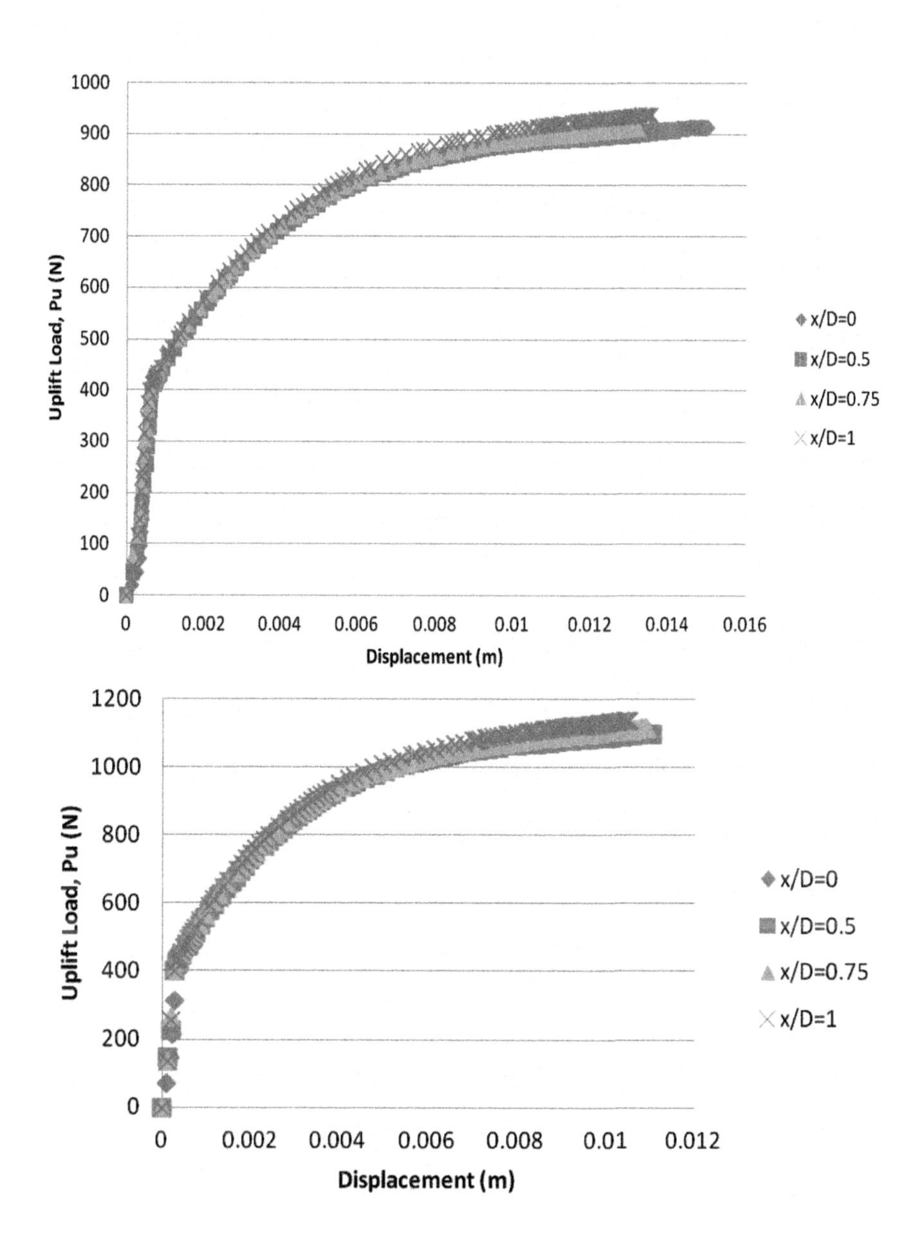
1000
900
800
700
600
500
400
300
200
100
0
0
0.002
0.004
0.006
0.008
0.01
0.012
0.014
0.016
Uplift Load, Pu (N)
Displacement (m)
x/D=0
x/D=0.5
x/D=0.75
x/D=1
1200
1000
800
600
400
200
0
0
0.002
0.004
0.006
0.008
0.01
0.012
Uplift Load, Pu (N)
Displacement (m)
x/D=0
x/D=0.5
x/D=0.75
x/D=1

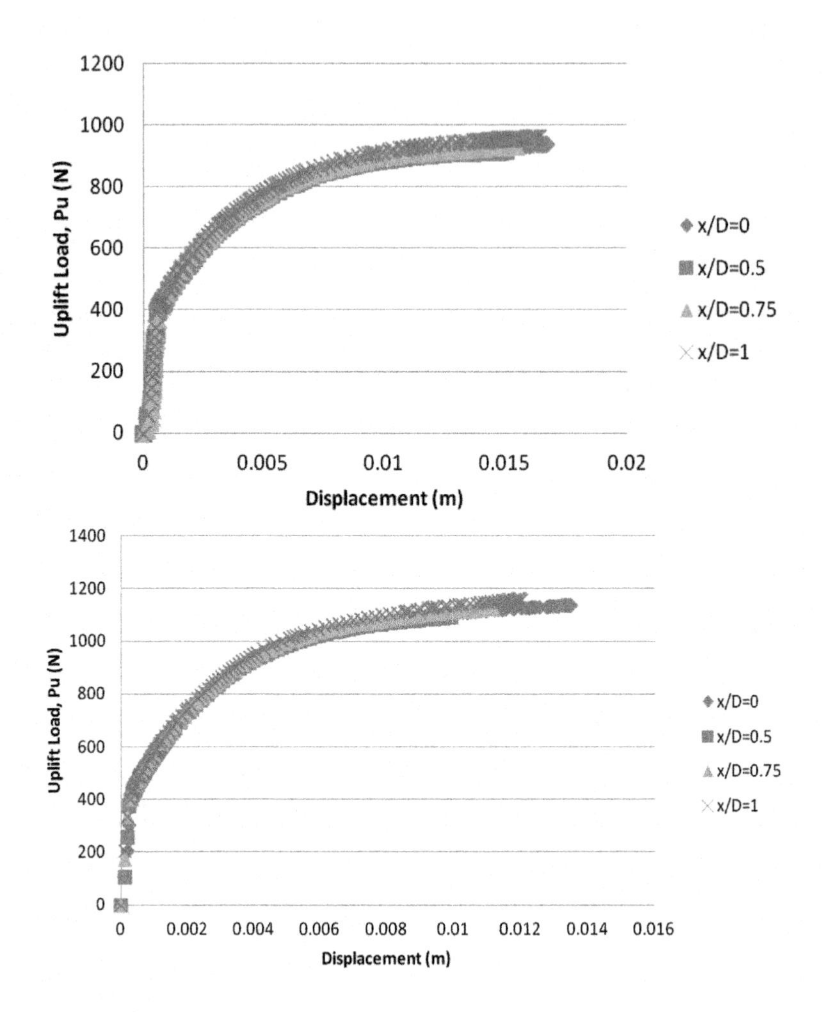
1200
1000
800
600
400
200
0
0
0.005
0.01
0.015
0.02
Uplift Load, Pu (N)
Displacement (m)
x/D=0
x/D=0.5
x/D=0.75
x/D=1
1400
1200
1000
800
600
400
200
0
0
0.002
0.004
0.006
0.008
0.01
0.012
0.014
0.016
Uplift Load, Pu (N)
Displacement (m)
x/D=0
x/D=0.5
x/D=0.75
x/D=1

APPENDIX C

Variation of Breakout Factor With Embedment Ratio Based on the Author's Empirical Formula

Models	Experimental results, N_q		PLAXIS results, N_q		Embedment ratio, L/D
	Loose sand	Dense sand	Loose sand	Dense sand	L/D
Circular plates	2.12	3.38	2.63	2.91	1
	5.77	8.98	7.05	7.78	2
	10.62	16.09	13	13.92	3
	13.66	20.73	16.74	17.88	4
Square plates	1.66	2.64	2.06	2.29	1
	4.53	6.93	5.53	5.99	2
	8.27	12.45	10.14	10.75	3
	10.67	15.98	13	13.80	4
Rectangular plates	5.53	8.98	6.80	7.75	1
	15.14	23.14	18.47	19.97	2
	29.15	44.59	35.55	38.48	3
	46.49	69.58	56.70	60.22	4

Friction angle in loose sand: Ø = 38 degrees.
Friction angle in dense sand: Ø = 44 degrees.

APPENDIX D

Variation of Breakout Factor With Embedment Ratio Based on Balla's Theory (1961) in Loose Sand

[1]	[2]	[3]	[4]	[5]	[6]	[7]	[8]
Model size	**D (mm)**	**L/D**	**Depth (mm)**	**$F_1 + F_3$**	**Theoretical pullout load (N) in circular**	**Breakout factor, N_q, in circular**	**Breakout factor, N_q, in square**
Circular plates	100	1	100	2.42	36	3.07	
		2	200	1.25	150	6.41	
		3	300	0.9	364	10.37	
		4	400	0.78	748	15.98	
Square plates	100	1	100	2.42	36		2.88
		2	200	1.25	150		5.62
		3	300	0.9	364		8.86
		4	400	0.78	748		13.31

Friction angle in loose sand: Ø = 38 degrees
Unit weight: 15 kN/m^3
$F_1 + F_3$: Fig. 2.39 in Chapter 2, Literature Review.
Theoretical pullout load (N) in circular:

$$\gamma(F_1 + F_3)H^3$$

Breakout factor, N_q in circular:

$$\frac{P_u}{\gamma \mathrm{AH}}$$

Breakout factor, N_q, in square:

$$N_q(\text{square}) * \text{correlation factor}$$

APPENDIX E

Variation of Breakout Factor With Embedment Ratio Based on Balla's Theory (1961) in Dense Sand

[1]	[2]	[3]	[4]	[5]	[6]	[7]	[8]
Model size	***D* (mm)**	***L/D***	**Depth (mm)**	**$F_1 + F_3$**	**Theoretical pullout load (N) in circular**	**Breakout factor, N_q, in circular**	**Breakout factor, N_q, in square**
Circular plates	100	1	100	2.5	42.5	3.19	
		2	200	1.3	177	6.65	
		3	300	0.95	436.5	10.93	
		4	400	0.8	871.4	16.37	
Square plates	100	1	100	2.5	42.5		2.92
		2	200	1.3	177		5.83
		3	300	0.95	436.5		9.34
		4	400	0.8	871.4		13.64

Friction angle in loose sand: Ø = 44 degrees
Unit weight: 17 kN/m^3
$F_1 + F_3$: Fig. 2.39 in Chapter 2, Literature Review.
Theoretical pullout load (N) in circular:

$$\gamma(F_1 + F_3)H^3$$

Breakout factor, N_q, in circular:

$$\frac{P_u}{\gamma \mathrm{AH}}$$

Breakout factor, N_q, in square:

$$N_q(\text{square}) * \text{correlation factor}$$

APPENDIX F

Variation of Breakout Factor With Embedment Ratio Based on the Meyerhof and Adams Theory (1968) in Loose Sand

[1]	[2]	[3]	[4]	[5]	[6]	[7]	[8]
Model size	**D (mm)**	**L/D**	**Depth (mm)**	**Shape factor**	**$K_u$tan Ø**	**Theoretical pullout load, P_u (N)**	**Breakout factor, N_q**
Square plates	100	1	100	1.32	0.73	43.89	2.92
		2	200	1.64	0.73	173.55	5.78
		3	300	1.96	0.73	431	9.57
		4	400	2.28	0.73	858	14.3
Rectangular plates	200	1	200	1.32	0.73	192.54	6.41
		2	400	1.64	0.73	945.5	15.75
		3	600	1.96	0.73	2581	28.67
		4	800	2.28	0.73	5454	45.45

Friction angle in loose sand: Ø = 38 degrees
Unit weight: 15 kN/m^3
Shape factor: m was shown Fig. 2.41 in Chapter 2, Literature Review

$$S_f = 1 + m\frac{L}{D}$$

K_u : Fig. 2.40 in Chapter 2, Literature Review.
Theoretical pullout load in rectangular plates, P_u (N):

$$P_u = W + \gamma H^2(2S_f L + B - L)K_u \tan Ø$$

Theoretical pullout load in square plates, P_u (N):

$$= + 2^2$$

Breakout factor, N_q:

$$N_q = 1 + \frac{L}{D} \quad K_u \, tan \varnothing$$

APPENDIX G

Variation of Breakout Factor With Embedment Ratio Based on the Meyerhof and Adams Theory (1968) in Dense Sand

[1]	[2]	[3]	[4]	[5]	[6]	[7]	[8]
Model size	***D* (mm)**	***L/D***	**Depth (mm)**	**Shape factor**	**$K_u$tan Ø**	**Theoretical pullout load, P_u (N)**	**Breakout factor, N_q**
Square plates	100	1	100	1.48	0.92	63	3.70
		2	200	1.96	0.92	279	8.20
		3	300	2.44	0.92	738	14.47
		4	400	2.92	0.92	1531	22.51
Rectangular plates	200	1	200	1.48	0.92	308	9.05
		2	400	1.96	0.92	1653	24.30
		3	600	2.44	0.92	4667	45.75
		4	800	2.92	0.92	10337	76

Friction angle in loose sand: Ø = 44 degrees
Unit weight: 17 kN/m^3
Shape factor: m was shown Fig. 2.41 in Chapter 2, Literature Review.

$$S_f = 1 + m\frac{L}{D}$$

K_u: Fig. 2.40 in Chapter 2, Literature Review
Theoretical pullout load in rectangular plates, P_u (N):

$$P_u = W + \gamma H^2(2S_fL + B - L)K_u \tan Ø$$

Theoretical pullout load in square plates, P_u (N):

$$= + 2^2$$

Breakout factor, N_q:

$$N_q = 1 + \frac{L}{D} K_u \tan \emptyset$$

APPENDIX H

Variation of Breakout Factor With Embedment Ratio Based on Vesic's Theory (1971) in Loose Sand

[1]	[2]	[3]	[4]	[5]	[6]	[7]
Model size	***D*** **(mm)**	***L/D***	**Depth (mm)**	**Breakout factor, N_q (rectangular)**	**Correlation factor**	**Breakout factor, N_q (square)**
Square plates	100	1	100	1.6	1.93	3.08
		2	200	2.2	2.63	5.78
		3	300	2.7	2.87	7.74
		4	400	3.3	2.98	9.83
Rectangular plates	200	1	200	1.6	–	–
		2	400	2.2	–	–
		3	600	2.7	–	–
		4	800	3.3	–	–

Friction angle in loose sand: Ø = 35 degrees
Unit weight: 15 kN/m^3
Breakout factor, N_q

$$N_q = \left[1 + A_1\left(\frac{H}{h_1/2}\right) + A_2\left(\frac{H}{h_1/2}\right)^2\right]$$

Theoretical pullout load, P_u (N):

$$P_u = \gamma H A N_q$$

APPENDIX I

Variation of Breakout Factor With Embedment Ratio Based on Vesic's Theory (1971) in Dense Sand

[1]	[2]	[3]	[4]	[5]	[6]	[7]
Model size	***D* (mm)**	***L/D***	**Depth (mm)**	**Breakout factor, N_q (rectangular)**	**Correlation factor**	**Breakout factor, N_q (square)**
Square plates	100	1	100	1.7	1.93	3.28
		2	200	2.3	2.63	6.04
		3	300	2.92	2.87	8.38
		4	400	3.50	2.98	10.43
Rectangular plates	200	1	200	1.7	–	–
		2	400	2.3	–	–
		3	600	2.92	–	–
		4	800	3.50	–	–

Friction angle in loose sand: Ø = 44 degrees
Unit weight: 17 kN/m^3
Breakout factor, N_q

$$N_q = \left[1 + A_1\left(\frac{H}{h_1/2}\right) + A_2\left(\frac{H}{h_1/2}\right)^2\right]$$

Theoretical pullout load, P_u (N):

$$P_u = \gamma H A N_q$$

APPENDIX J

Variation of Breakout Factor With Embedment Ratio Based on the Rowe and Davis Theory (1982) in Loose Sand

[1]	[2]	[3]	[4]	[5]
Model size	***D* (mm)**	***L/D***	**Depth (mm)**	**Breakout factor, N_q**
Circular plates	100	1	100	1.6
		2	200	2.35
		3	300	3
		4	400	3.8

Friction angle in loose sand: Ø = 38 degrees
Unit weight: 15 kN/m^3
Breakout factor, N_q Fig. 2.45 in Chapter 2, Literature Review.

APPENDIX K

Variation of Breakout Factor With Embedment Ratio Based on the Rowe and Davis Theory (1982) in Dense Sand

[1]	[2]	[3]	[4]	[5]
Model size	***D* (mm)**	***L/D***	**Depth (mm)**	**Breakout factor, N_q**
Circular plates	100	1	100	1.8
		2	200	2.6
		3	300	3.15
		4	400	4.10

Friction angle in dense sand: Ø = 44 degrees
Unit weight: 17 kN/m^3
Breakout factor, N_q Fig. 2.45 in Chapter 2, Literature Review.

APPENDIX L

Variation of Breakout Factor With Embedment Ratio Based on the Murray and Geddes Theory (1987) in Loose Sand

[1]	[2]	[3]	[4]	[5]
Model size	***D*** **(mm)**	***L/D***	**Depth (mm)**	**Breakout factor,** N_q
Square plates	100	1	100	3.37
		2	200	7.36
		3	300	12.98
		4	400	20.21
Rectangular plates	200	1	200	8.14
		2	400	21.77
		3	600	41.90
		4	800	68.51

Friction angle in loose sand: $\varnothing = 38$ degrees
Unit weight: 15 kN/m^3
Breakout factor, N_q

$$N_q = 1 + \frac{L}{D}\tan\varnothing\left(1 + \frac{D}{B} + \frac{\pi}{3}\frac{L}{B}\right)$$

APPENDIX M

Variation of Breakout Factor With Embedment Ratio Based on the Murray and Geddes Theory (1987) in Dense Sand

[1]	[2]	[3]	[4]	[5]
Model size	***D* (mm)**	***L/D***	**Depth (mm)**	**Breakout factor, N_q**
Square plates	100	1	100	3.91
		2	200	8.83
		3	300	15.74
		4	400	24.65
Rectangular plates	200	1	200	9.79
		2	400	26.57
		3	600	51.34
		4	800	84.09

Friction angle in loose sand: Ø = 44 degrees
Unit weight: 17 kN/m^3
Breakout factor, N_q:

$$N_q = 1 + \frac{L}{D} tan\varnothing\left(1 + \frac{D}{B} + \frac{\pi}{3}\frac{L}{B}\right)$$

APPENDIX N

Variation of Breakout Factor With Embedment Ratio Based on the Dickin and Laman Findings (2007) in Loose Sand

[1]	[2]	[3]	[4]	[5]
Model size	**D (mm)**	**L/D**	**Depth (mm)**	**Breakout factor, N_q**
Rectangular plates	200	1	200	1.25
		2	400	1.88
		3	600	2.35
		4	800	2.90

Friction angle in loose sand: Ø = 38 degrees
Unit weight: 15 kN/m^3
Breakout factor, N_q: Fig. 2.49 in Chapter 2, Literature Review.

APPENDIX O

Variation of Breakout Factor With Embedment Ratio Based on the Dickin and Laman Findings (2007) in Dense Sand

[1]	[2]	[3]	[4]	[5]
Model size	***D* (mm)**	***L/D***	**Depth (mm)**	**Breakout factor, N_q**
Rectangular plates	200	1	200	1.90
		2	400	2.94
		3	600	3.86
		4	800	5.10

Friction angle in loose sand: Ø = 44 degrees
Unit weight: 17 kN/m^3
Breakout factor, N_q: Fig. 2.50 in Chapter 2, Literature Review.

APPENDIX P

Regression and Analysis

Regression and Analysis

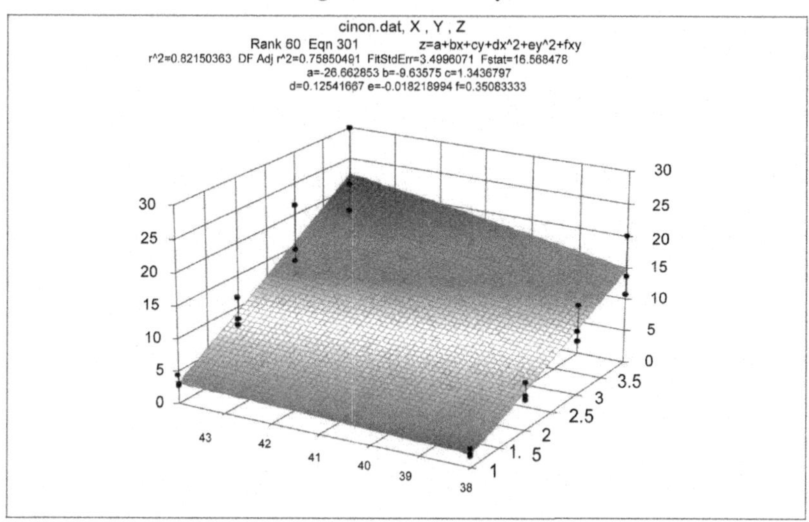

Rank 1 Eqn 314 $z=a+b\ln x+c\ln y+d(\ln x)^2+e(\ln y)^2+f\ln x\ln y+g(\ln x)^3+h(\ln y)^3+i\ln x(\ln y)^2+j(\ln x)^2\ln y$

r^2 Coef Det	DF Adj r^2	Fit Std Err	F-value
0.8232424227	0.6872750555	3.9488064508	7.2449472522

Parm	Value	Std Error	t-value	95.00% Confidence Limits		P>|t|
a	-240.967343	5.71067e+07	-4.2196e-06	-1.2248e+08	1.22482e+08	1.00000
b	568.0879872	6.90915e+07	8.22226e-06	-1.4819e+08	1.48187e+08	0.99999
c	77.46925904	4.53689e+07	1.70754e-06	-9.7307e+07	9.73067e+07	1.00000
d	-44.1774084	201.0195625	-0.21976671	-475.32149	386.9666735	0.82922
e	9.116124856	8.40812e+06	1.08421e-06	-1.8034e+07	1.80336e+07	1.00000
f	-316.333937	3.72517e+07	-8.4918e-06	-7.9897e+07	7.98966e+07	0.99999
g	-3.07405265	16.21967132	-0.1895262	-37.8617878	31.71368251	0.85240
h	-3.31210687					
i	43.89106895	5.01924e+06	8.74456e-06	-1.0765e+07	1.07653e+07	0.99999
j	15.49980119	53.33265561	0.290624965	-98.8873687	129.8869711	0.77560

X at Fn Zmin	Y at Fn Zmin	Fn Zmin
1	38	2.0381117081
X at Fn Zmax	Y at Fn Zmax	Fn Zmax
4	44	22.252185051

Procedure
GaussElim

r^2 Coef Det	DF Adj r^2	Fit Std Err	r^2 Attainable
0.8232424227	0.6872750555	3.9488064508	0.8232435671

Source	Sum of Squares	DF	Mean Square	F Statistic	P>F
Regr	1016.7389	9	112.97099	7.24495	0.00061
Error	218.30301	14	15.593072		
Total	1235.0419	23			
Lack Fit	0.0014134037	-2	-0.00070670185	-5.17964e-05	0.00000
Pure Err	218.3016	16	13.64385		

Description: cinon.dat, X , Y , Z

X Variable:

Xmin:	1	Xmax:	4	Xrange:	3
Xmean:	2.5	Xstd:	1.1420804814		

Y Variable:

Ymin:	38	Ymax:	44	Yrange:	6
Ymean:	41	Ystd:	3.0645235107		

Z Variable:

Zmin:	1.6	Zmax:	29.88	Zrange:	28.28
Zmean:	10.449583333	Zstd:	7.327856012		

Date	Time	File Source
Apr 5, 2012	3:13:02 PM	d:\program files\tablecurve3dv4.0\cinon

cinon.dat, X ; Y , Z

Rank 1 Eqn 314 z=a+blnx+clny+d(lnx)^2+e(lny)^2+flnxlny+g(lnx)^3+h(lny)^3+ilnx(lny)^2+j(lnx)^2lny

r^2=0.82324242 DF Adj r^2=0.68727506 FitStdErr=3.9488065 Fstat=7.2449473 a=-240.96734 b=568.08799 c=77.469259 d=-44.177408 e=9.1161249 f=-316.33394 g=-3.0740526 h=-3.3121069 i=43.891069 j=15.499801

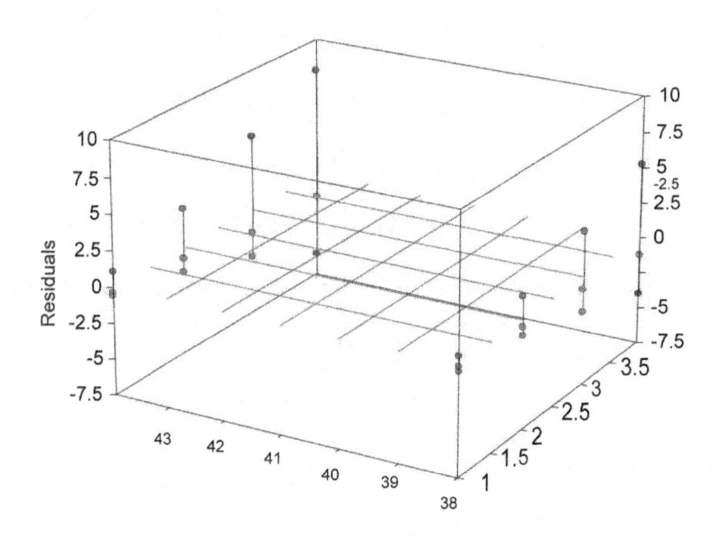

Rank 1 Eqn 314 $z=a+b\ln x+c\ln y+d(\ln x)^2+e(\ln y)^2+f\ln x\ln y+g(\ln x)^3+h(\ln y)^3+i\ln x(\ln y)^2+j(\ln x)^2\ln y$

Precision	Avg Abs Error	Min Abs Error	Max Abs Error
18	7.201602e-15	0	1.394512e-14
17	1.066812e-14	0	1.847094e-14
16	1.44079e-14	8.740063e-15	2.31105e-14
15	1.627391e-13	6.992051e-14	2.484133e-13
14	1.105343e-12	4.282631e-13	1.876901e-12
13	5.82571e-12	3.765347e-12	8.896986e-12
12	7.765327e-11	4.862097e-11	1.030294e-10
11	8.540786e-11	2.513637e-12	2.000288e-10
10	2.668274e-09	1.367156e-09	3.828063e-09
9	4.765954e-08	2.388478e-08	9.946658e-08
8	6.302467e-07	4.338042e-07	1.12262e-06
7	1.633146e-05	1.175243e-05	2.241871e-05
6	6.032056e-05	2.592469e-05	0.000136097
5	0.0012193064	0.0009238887	0.0017649715
4	0.0087674562	0.0064812424	0.0129872961
3	0.113194629	0.0779044268	0.1647675
2	0.7432678184	0.1178579049	1.3249177268

Removing	Avg Abs Error	Min Abs Error	Max Abs Error
a	43.279313838	10.828929507	118.23068497
b	31.740767039	0	63.188940985
c	51.488347368	13.174363162	138.26577973
d	2.7953561904	0	5.7529299806
e	22.42647307	5.8665582234	59.184619315
f	65.422432082	0	127.99250716
g	0.2297088775	0	0.5549522424
h	30.16757308	8.065853947	78.219714072
i	33.608458185	0	64.599358722
j	3.6303384066	0	7.3422348105

```
/* ---------------------------------------------------------------*
             To modify generated output, edit C.TCL
                                                                */

#include <math.h>
#include <stdio.h>

double eqn314(double x, double y);

void main(void)
{
  double  x,y,z;
  char  str[80];
  while(1){
    printf("Enter x:
    "); gets(str);
    if(!*str) break;
    sscanf(str,"%lg",&x)
    ; printf("Enter y:
    "); gets(str);
    if(!*str) break;
    sscanf(str,"%lg",&y)
    ;
    z=eqn314(x,y);
    printf("      z=%.15lg",z);
    }
}
```

```
/*--------------------------------------------------------------*/
double eqn314(double x, double y)
/*--------------------------------------------------------------*

   File Source= d:filesdv4.0.dat
   Date= Apr 5, 2012
   Time= 3:13:04 PM
   Data Set= cinon.dat, X , Y ,
   Z X=
   Y=
   Z=
   Eqn#=
   314 Eqn=
z=a+blnx+clny+d(lnx)^2+e(lny)^2+flnxlny+g(lnx)^3+h(lny)^3+ilnx(lny)^2+j
(lnx)^2lny
   r2=0.8232424226739305
   r2adj=0.6872750555000308
   StdErr=3.948806450812484
   Fstat=7.24494725223061
   2 a= -
   240.9673432996002 b=
   568.0879871843416
   c= 77.4692590421687 d=
   -44.17740840887954 e=
   9.116124856210058 f= -
   316.3339369133846 g= -
   3.074052645265573 h= -
   3.31210687013229 i=
   43.89106895249982 j=
   15.4998011872754

 -------------------------------------------------------------*/
{
  double z;
  x=log(x);
  y=log(y);
  z=-240.9673432996002+
    x*(568.0879871843416+x*(-44.17740840887954+
    x*(-3.074052645265573)))+y*(77.46925904216870+
    y*(9.116124856210058+y*(-3.312106870132290)))+
    x*y*(-316.3339369133846+y*(43.89106895249982)+
    x*(15.49980118727540));
  return z;
}
```

1 0.8232424227 33 **314**
$z=a+b\ln x+c\ln y+d(\ln x)^2+e(\ln y)^2+f\ln x\ln y+g(\ln x)^3+h(\ln y)^3+i\ln x(\ln y)^2+j(\ln x)^2\ln y$

2 0.8232424227 29 **317**
$z=a+b\ln x+c/y+d(\ln x)^2+e/y^2+f(\ln x)/y+g(\ln x)^3+h/y^3+i(\ln x)/y^2+j(\ln x)^2/y$

3 0.8232424227 28 **311** $z=a+b\ln x+cy+d(\ln x)^2+ey^2+fy\ln x+g(\ln x)^3+hy^3+iy^2\ln x+jy(\ln x)^2$

4 0.8232293807 24 **316** $z=a+bx+c/y+dx^2+e/y^2+fx/y+gx^3+h/y^3+ix/y^2+jx^2/y$

5 0.8232293807 23 **310** $z=a+bx+cy+dx^2+ey^2+fxy+gx^3+hy^3+ixy^2+jx^2y$

6 0.8232293807 28 **313** $z=a+bx+c\ln y+dx^2+e(\ln y)^2+fx\ln y+gx^3+h(\ln y)^3+ix(\ln y)^2+jx^2\ln y$

7 0.8232084803 229 **2108**
z=a+LNCUMX(b,c,d)+LNCUMY(e,f,g)+LNCUMX(h,c,d)*LNCUMY(1,f,g)

8 0.8232084803 74 **2132**
z=a+EXVCUMX(b,c,d)+EXVCUMY(e,f,g)+EXVCUMX(h,c,d)*EXVCUMY(1,f,g)

9 0.8232084803 49 **2095**
z=LORCUMX(a,b,c)+LORCUMY(d,e,f)+LORCUMX(g,b,c)*LORCUMY(1,e,f)

10 0.8232084803 32 **2024** z=a+LORX(b,c,d)+LORY(e,f,g)+LORX(h,c,d)*LORY(1,f,g)

11 0.8232084803 75 **2143** z=LDRX(a,b,c)+LDRY(d,e,f)+LDRX(g,b,c)*LDRY(1,e,f)

12 0.8232084803 106 **2048**
z=a+LOGISTICX(b,c,d)+LOGISTICY(e,f,g)+LOGISTICX(h,c,d)*LOGISTICY(1,f,g)

13 0.8232084803 108 **2036**
z=a+LOGNORMX(b,c,d)+LOGNORMY(e,f,g)+LOGNORMX(h,c,d)*LOGNORMY(1,f,g)

14 0.8232084803 228 **2107**
z=LNCUMX(a,b,c)+LNCUMY(d,e,f)+LNCUMX(g,b,c)*LNCUMY(1,e,f)

15 0.8232084803 50 **2096**
z=a+LORCUMX(b,c,d)+LORCUMY(e,f,g)+LORCUMX(h,c,d)*LORCUMY(1,f,g)

16 0.8232084803 204 **2084**
z=a+GCUMX(b,c,d)+GCUMY(e,f,g)+GCUMX(h,c,d)*GCUMY(1,f,g)

17 0.8232084803 107 **2035**
z=LOGNORMX(a,b,c)+LOGNORMY(d,e,f)+LOGNORMX(g,b,c)*LOGNORMY(1,e,f)

18 0.8232084803 84 **2071**
z=a+GAUSSX(b,c,d)+LOGNORMY(e,f,g)+GAUSSX(h,c,d)*LOGNORMY(1,f,g)

19 0.8232084803 52 **2120** z=a+SIGX(b,c,d)+SIGY(e,f,g)+SIGX(h,c,d)*SIGY(1,f,g)

20 0.8232084803 73 **2131**
z=EXVCUMX(a,b,c)+EXVCUMY(d,e,f)+EXVCUMX(g,b,c)*EXVCUMY(1,e,f)

21 0.8232084803 76 **2144** z=a+LDRX(b,c,d)+LDRY(e,f,g)+LDRX(h,c,d)*LDRY(1,f,g)

22 0.8232070918 83 **2069**
z=GAUSSX(a,b,c)+LOGNORMY(d,e,f)+GAUSSX(g,b,c)*LOGNORMY(1,e,f)

23 0.8231993556 41 **2063** z=a+GAUSSX(b,c,d)*LOGNORMY(1,e,f)

24 0.8231943746 102 **2074** z=a+GCUMX(b,c,d)*GCUMY(1,e,f)

25 0.8231879082 26 **2110** z=a+SIGX(b,c,d)*SIGY(1,e,f)

26 0.8231879082 26 **2112** z=a+SIGX(b,c,d)*SIGY(1,e,d)

27 0.8231849888 203 **2083**
z=GCUMX(a,b,c)+GCUMY(d,e,f)+GCUMX(g,b,c)*GCUMY(1,e,f)

28 0.823155213 29 **315**
$z=a+b/x+c\ln y+d/x^2+e(\ln y)^2+f(\ln y)/x+g/x^3+h(\ln y)^3+i(\ln y)^2/x+j(\ln y)/x^2$

29 0.823155213 24 **312** $z=a+b/x+cy+d/x^2+ey^2+fy/x+g/x^3+hy^3+iy^2/x+jy/x^2$

30 0.823155213 25 **318** $z=a+b/x+c/y+d/x^2+e/y^2+f/(xy)+g/x^3+h/y^3+i/(xy^2)+j/(x^2y)$

31 0.8231140586 101 **2073** z=GCUMX(a,b,c)*GCUMY(1,d,e)

32 0.8230098707 59 **2011**
z=GAUSSX(a,b,c)+GAUSSY(d,e,f)+GAUSSX(g,b,c)*GAUSSY(1,e,f)

33 0.8229789046 40 **2061** z=GAUSSX(a,b,c)*LOGNORMY(1,d,e)

34 0.8229673549 36 **2122** z=a+EXVCUMX(b,c,d)*EXVCUMY(1,e,f)

35 0.8229627184 37 **2123** z=EXVCUMX(a,b,c)*EXVCUMY(1,d,c)

36 0.8229627184 36 **2121** z=EXVCUMX(a,b,c)*EXVCUMY(1,d,e)

37 0.8229453777 44 **2040** z=a+LOGISTICX(b,c,d)*LOGISTICY(1,e,d)

38 0.8229453777 44 **2038** z=a+LOGISTICX(b,c,d)*LOGISTICY(1,e,f)

39 0.8229364899 51 **2119** z=SIGX(a,b,c)+SIGY(d,e,f)+SIGX(g,b,c)*SIGY(1,e,f)

40 0.8229132343 25 **2109** z=SIGX(a,b,c)*SIGY(1,d,e)

41 0.8225115925 38 **2134** z=a+LDRX(b,c,d)*LDRY(1,e,f)

42 0.8224186279 37 **2133** z=LDRX(a,b,c)*LDRY(1,d,e)

43 0.8221551545 45 **2026** z=a+LOGNORMX(b,c,d)*LOGNORMY(1,e,f)

44 0.8221149783 83 **2070**
z=LOGNORMX(a,b,c)+GAUSSY(d,e,f)+LOGNORMX(g,b,c)*GAUSSY(1,e,f)

45	0.8221123697	105	**2047**	z=LOGISTICX(a,b,c)+LOGISTICY(d,e,f)+LOGISTICX(g,b,c)*LOGISTICY(1,e,f)
46	0.8221101652	43	**2037**	z=LOGISTICX(a,b,c)*LOGISTICY(1,d,e)
47	0.8221101652	43	**2039**	z=LOGISTICX(a,b,c)*LOGISTICY(1,d,c)
48	0.8221058739	115	**2098**	z=a+LNCUMX(b,c,d)*LNCUMY(1,e,f)
49	0.8220951185	45	**2028**	z=a+LOGNORMX(b,c,d)*LOGNORMY(1,e,d)
50	0.8220849099	44	**2025**	z=LOGNORMX(a,b,c)*LOGNORMY(1,d,e)
51	0.8220566731	114	**2097**	z=LNCUMX(a,b,c)*LNCUMY(1,d,e)
52	0.8219088065	115	**2100**	z=a+LNCUMX(b,c,d)*LNCUMY(1,e,d)
53	0.8218079134	44	**2027**	z=LOGNORMX(a,b,c)*LOGNORMY(1,d,c)
54	0.8217225215	21	**305**	$z=a+blnx+clny+d(lnx)^{2}+e(lny)^{2}+flnxlny$
55	0.8217225215	17	**308**	$z=a+blnx+c/y+d(lnx)^{2}+e/y^{2}+f(lnx)/y$
56	0.8217225215	16	**302**	$z=a+blnx+cy+d(lnx)^{2}+ey^{2}+fylnx$
57	0.821614104	60	**2168**	z=a+POWX(b,c)+POWY(d,e)+POWX(f,c)*POWY(1,e)
58	0.8215222706	25	**2086**	z=a+LORCUMX(b,c,d)*LORCUMY(1,e,f)
59	0.8215222706	25	**2088**	z=a+LORCUMX(b,c,d)*LORCUMY(1,e,d)
60	0.8215036322	11	**301**	$z=a+bx+cy+dx^{2}+ey^{2}+fxy$
61	0.8215036322	12	**307**	$z=a+bx+c/y+dx^{2}+e/y^{2}+fx/y$
62	0.8215036322	16	**304**	$z=a+bx+clny+dx^{2}+e(lny)^{2}+fxlny$
63	0.821486469	30	**2158**	z=a+POWX(b,c)*POWY(1,d)
64	0.8214739582	42	**2156**	z=a+EXPX(b,c)+EXPY(d,e)+EXPX(f,c)*EXPY(1,e)
65	0.8214261826	13	**2016**	z=a+LORX(b,c,d)*LORY(1,e,d)
66	0.8214261826	13	**2014**	z=a+LORX(b,c,d)*LORY(1,e,f)
67	0.8210976525	31	**2023**	z=LORX(a,b,c)+LORY(d,e,f)+LORX(g,b,c)*LORY(1,e,f)
68	0.8208064474	38	**2136**	z=a+LDRX(b,c,d)*LDRY(1,e,d)
69	0.820530257	59	**2167**	z=POWX(a,b)+POWY(c,d)+POWX(e,b)*POWY(1,d)
70	0.8201630521	29	**2157**	z=POWX(a,b)*POWY(1,c)
71	0.8185552915	114	**2099**	z=LNCUMX(a,b,c)*LNCUMY(1,d,c)
72	0.818084966	17	**306**	$z=a+b/x+clny+d/x^{2}+e(lny)^{2}+f(lny)/x$
73	0.818084966	12	**303**	$z=a+b/x+cy+d/x^{2}+ey^{2}+fy/x$
74	0.818084966	13	**309**	$z=a+b/x+c/y+d/x^{2}+e/y^{2}+f/(xy)$
75	0.8173624115	12	**2013**	z=LORX(a,b,c)*LORY(1,d,e)
76	0.8173624115	12	**2015**	z=LORX(a,b,c)*LORY(1,d,c)
77	0.816481559	24	**2085**	z=LORCUMX(a,b,c)*LORCUMY(1,d,e)
78	0.8158527675	37	**2135**	z=LDRX(a,b,c)*LDRY(1,d,c)
79	0.8151924969	41	**2155**	z=EXPX(a,b)+EXPY(c,d)+EXPX(e,b)*EXPY(1,d)
80	0.809340661	21	**2146**	z=a+EXPX(b,c)*EXPY(1,d)
81	0.7985621312	21	**151237000**	$lnz=a+bx^{2.5}+cx^{0.5}+d/y$
82	0.7985621312	24	**151236998**	$lnz=a+bx^{2.5}+cx^{0.5}+d/y^{0.5}$
83	0.7985621312	28	**151237002**	$lnz=a+bx^{2.5}+cx^{0.5}+dlny/y^{2}$
84	0.7985621312	28	**151236991**	$lnz=a+bx^{2.5}+cx^{0.5}+de^{y/wy}$
85	0.7985621312	22	**151236990**	$lnz=a+bx^{2.5}+cx^{0.5}+dy^{3}$
86	0.7985621312	27	**151237004**	$lnz=a+bx^{2.5}+cx^{0.5}+de^{-y}$
87	0.7985621312	23	**151236986**	$lnz=a+bx^{2.5}+cx^{0.5}+dy^{1.5}$
88	0.7985621312	26	**151236985**	$lnz=a+bx^{2.5}+cx^{0.5}+dylny$
89	0.7985621312	27	**151236999**	$lnz=a+bx^{2.5}+cx^{0.5}+dlny/y$
90	0.7985621312	27	**151236997**	$lnz=a+bx^{2.5}+cx^{0.5}+d/lny$
91	0.7985610165	47	**152041716**	$lnz=a+bx^{0.5}+clnx+de^{-x}+eylny+fe^{y/wy}$
92	0.7985610165	33	**152041706**	$lnz=a+bx^{0.5}+clnx+de^{-x}+ey+f/y$
93	0.7985610165	43	**151948332**	$lnz=a+bx^{0.5}lnx+cx^{0.5}+dlnx+elny/y+f/y$
94	0.7985610165	44	**151953056**	$lnz=a+bx^{0.5}lnx+cx^{0.5}+de^{-x}+elny/y+f/y$
95	0.7985610165	47	**151960996**	$lnz=a+bx^{0.5}lnx+clnx+de^{-x}+elny/y+f/y$
96	0.7985610165	40	**152041886**	$lnz=a+bx^{0.5}+clnx+de^{-x}+elny/y+f/y$
97	0.7985610165	37	**152075578**	$lnz=a+blnx+c/x^{0.5}+d/x+elny/y+f/y$
98	0.7985610165	53	**203853410**	$lnz=a+bx+cx^{0.5}lnx+d/x^{0.5}+e/x+fy+ge^{y/wy}+hlny+i/y$
99	0.7985610165	40	**152041725**	$lnz=a+bx^{0.5}+clnx+de^{-x}+eylny+f/y$
100	0.7985610165	41	**151949721**	$lnz=a+bx^{0.5}lnx+cx^{0.5}+d/x^{0.5}+elny/y+f/y$
101	0.7985610165	42	**152041818**	$lnz=a+bx^{0.5}+clnx+de^{-x}+ee^{y/wy}+f/y$
102	0.7985610165	53	**156286868**	$lnz=a+bx^{0.5}lnx+cx^{0.5}+dlnx+eylny+fe^{y/wy}+g/y$
103	0.7985610165	47	**156962961**	$lnz=a+blnx+c/x^{0.5}+d/x+eylny+fe^{y/wy}+g/y$
104	0.7985610165	51	**156962957**	$lnz=a+blnx+c/x^{0.5}+d/x+eylny+fe^{y/wy}+glny$
105	0.7985610165	54	**183174651**	$lnz=a+blnx+c/x^{0.5}+d/x+eylny+fe^{y/wy}+glny+h/y$

106	0.7985610165	57	**156286864**	$lnz=a+bx^{0.5}lnx+cx^{0.5}+dlnx+eylny+fe^{y/wy}+glny$
107	0.7985610165	53	**183165726**	$lnz=a+blnx+c/x^{0.5}+d/x+ey+fylny+ge^{y/wy}+hlny$
108	0.7985610165	56	**301276669**	$lnz=a+blnx+c/x^{0.5}+d/x+ey+fylny+ge^{y/wy}+hlny+i/y$
109	0.7985610165	60	**180343312**	$lnz=a+bx^{0.5}lnx+cx^{0.5}+dlnx+eylny+fe^{y/wy}+glny+h/y$
110	0.7985610165	55	**180339766**	$lnz=a+bx^{0.5}lnx+cx^{0.5}+dlnx+ey+fe^{y/wy}+glny/y+h/y$
111	0.7985610165	55	**180336315**	$lnz=a+bx^{0.5}lnx+cx^{0.5}+dlnx+ey+fylny+ge^{y/wy}+h/y$
112	0.7985610165	45	**156286520**	$lnz=a+bx^{0.5}lnx+cx^{0.5}+dlnx+ey+flny/y+g/y$
113	0.7985610165	44	**156286359**	$lnz=a+bx^{0.5}lnx+cx^{0.5}+dlnx+ey+fylny+g/y$
114	0.7985610165	51	**156286936**	$lnz=a+bx^{0.5}lnx+cx^{0.5}+dlnx+eylny+flny/y+g/y$
115	0.7985610165	64	**291814995**	$lnz=a+bx^{0.5}lnx+cx^{0.5}+dlnx+ey+fylny+ge^{y/wy}+hlny/y+i/y$
116	0.7985610165	62	**180343330**	$lnz=a+bx^{0.5}lnx+cx^{0.5}+dlnx+eylny+fe^{y/wy}+glny/y+h/y$
117	0.7985610165	53	**180336383**	$lnz=a+bx^{0.5}lnx+cx^{0.5}+dlnx+ey+fylny+glny/y+h/y$
118	0.7985610165	53	**156288879**	$lnz=a+bx^{0.5}lnx+cx^{0.5}+dlnx+ee^{y/wy}+flny/y+g/y$
119	0.7985610165	46	**156286452**	$lnz=a+bx^{0.5}lnx+cx^{0.5}+dlnx+ey+fe^{y/wy}+g/y$
120	0.7985610165	59	**180339747**	$lnz=a+bx^{0.5}lnx+cx^{0.5}+dlnx+ey+fe^{y/wy}+glny+hlny/y$
121	0.7985610165	62	**291814977**	$lnz=a+bx^{0.5}lnx+cx^{0.5}+dlnx+ey+fylny+ge^{y/wy}+hlny+i/y$
122	0.7985610165	59	**180336311**	$lnz=a+bx^{0.5}lnx+cx^{0.5}+dlnx+ey+fylny+ge^{y/wy}+hlny$
123	0.7985610165	51	**156286451**	$lnz=a+bx^{0.5}lnx+cx^{0.5}+dlnx+ey+fe^{y/wy}+glny/y$
124	0.7985610165	37	**151325233**	$lnz=a+bx^{0.5}+clnx+de^{-x}+e/lny$
125	0.7985610165	35	**151325232**	$lnz=a+bx^{0.5}+clnx+de^{-x}+elny$
126	0.7985610165	31	**151325236**	$lnz=a+bx^{0.5}+clnx+de^{-x}+e/y$
127	0.7985610165	37	**151325235**	$lnz=a+bx^{0.5}+clnx+de^{-x}+elny/y$
128	0.7985610165	38	**151325227**	$lnz=a+bx^{0.5}+clnx+de^{-x}+ee^{y/wy}$
129	0.7985610165	59	**295258074**	$lnz=a+bx^{0.5}+clnx+d/x^{0.5}+e/x+fylny+ge^{y/wy}+hlny+i/y$
130	0.7985610165	68	**291814976**	$lnz=a+bx^{0.5}lnx+cx^{0.5}+dlnx+ey+fylny+ge^{y/wy}+hlny+ilny/y$
131	0.7985610165	66	**180343311**	$lnz=a+bx^{0.5}lnx+cx^{0.5}+dlnx+eylny+fe^{y/wy}+glny+hlny/y$
132	0.7985610165	60	**204518171**	$lnz=a+bx+cx^{0.5}+dlnx+e/x+fylny+ge^{y/wy}+hlny+ilny/y$
133	0.7985610165	33	**151311679**	$lnz=a+bx^{0.5}lnx+cx^{0.5}+dlnx+e/y$
134	0.7985610165	62	**292825269**	$lnz=a+bx^{0.5}lnx+cx^{0.5}+d/x^{0.5}+ey+fylny+ge^{y/wy}+hlny/y+i/y$
135	0.7985610165	69	**291874749**	$lnz=a+bx^{0.5}lnx+cx^{0.5}+dlnx+eylny+fe^{y/wy}+glny+hlny/y+i/y$
136	0.7985610165	39	**151311676**	$lnz=a+bx^{0.5}lnx+cx^{0.5}+dlnx+e/lny$
137	0.7985610165	40	**151311670**	$lnz=a+bx^{0.5}lnx+cx^{0.5}+dlnx+ee^{y/wy}$
138	0.7985610165	40	**151311872**	$lnz=a+bx^{0.5}lnx+cx^{0.5}+de^{-x}+e/lny$
139	0.7985610165	35	**151311875**	$lnz=a+bx^{0.5}lnx+cx^{0.5}+de^{-x}+e/y$
140	0.7985610165	63	**291852304**	$lnz=a+bx^{0.5}lnx+cx^{0.5}+dlnx+ey+fe^{y/wy}+glny+hlny/y+i/y$
141	0.7963206646	51	**7285441**	$z=a+bx+cx^{1.5}+dx^{0.5}lnx+ey^{1.5}+fy^{2}lny+g(lny)^{2}+he^{-y}$
142	0.7963206646	35	**149438**	$z=a+bx+cx^{2}+dx^{0.5}lnx+ey^{2}lny+fe^{-y}$
143	0.7963206646	40	**1115216**	$z=a+bx+cx^{1.5}+dx^{0.5}lnx+ey^{1.5}+fy^{2}lny+gy^{2.5}$
144	0.7963206646	32	**149392**	$z=a+bx+cx^{2}+dx^{0.5}lnx+ey^{1.5}+fe^{y/wy}$
145	0.7963206646	37	**140894**	$z=a+bx+cx^{1.5}+dx^{0.5}lnx+ey^{2}lny+fe^{-y}$
146	0.7963206646	49	**7285380**	$z=a+bx+cx^{1.5}+dx^{0.5}lnx+ey^{1.5}+fy^{2}lny+gy^{2.5}+h(lny)^{2}$
147	0.7963206646	38	**140936**	$z=a+bx+cx^{1.5}+dx^{0.5}lnx+ee^{y/wy}+fe^{-y}$
148	0.7963206646	36	**149480**	$z=a+bx+cx^{2}+dx^{0.5}lnx+ee^{y/wy}+fe^{-y}$
149	0.7963206646	34	**140848**	$z=a+bx+cx^{1.5}+dx^{0.5}lnx+ey^{1.5}+fe^{y/wy}$
150	0.7963206646	31	**149405**	$z=a+bx+cx^{2}+dx^{0.5}lnx+ey^{1.5}+fe^{-y}$
151	0.7963206646	52	**7287185**	$z=a+bx+cx^{1.5}+dx^{0.5}lnx+ey^{1.5}+fe^{y/wy}+g(lny)^{2}+he^{-y}$
152	0.7963206646	43	**1115262**	$z=a+bx+cx^{1.5}+dx^{0.5}lnx+ey^{1.5}+fe^{y/wy}+g(lny)^{2}$
153	0.7963206646	40	**1024604**	$z=a+bx+cxlnx+dx^{0.5}lnx+ex^{0.5}+fy^{1.5}+ge^{-y}$
154	0.7963206646	51	**7285378**	$z=a+bx+cx^{1.5}+dx^{0.5}lnx+ey^{1.5}+fy^{2}lny+gy^{2.5}+he^{y/wy}$
155	0.7963206646	49	**7285391**	$z=a+bx+cx^{1.5}+dx^{0.5}lnx+ey^{1.5}+fy^{2}lny+gy^{2.5}+he^{-y}$
156	0.7963206646	59	**39786066**	$z=a+bx+cx^{1.5}+dx^{0.5}lnx+ey^{1.5}+fy^{2}lny+gy^{2.5}+he^{y/wy}+i(lny)^{2}$
157	0.7963206646	58	**39786100**	$z=a+bx+cx^{1.5}+dx^{0.5}lnx+ey^{1.5}+fy^{2}lny+gy^{2.5}+h(lny)^{2}+ie^{-y}$
158	0.7963206646	33	**140861**	$z=a+bx+cx^{1.5}+dx^{0.5}lnx+ey^{1.5}+fe^{-y}$
159	0.7963206646	53	**7285418**	$z=a+bx+cx^{1.5}+dx^{0.5}lnx+ey^{1.5}+fy^{2}lny+ge^{y/wy}+he^{-y}$
160	0.7963206646	43	**1115273**	$z=a+bx+cx^{1.5}+dx^{0.5}lnx+ey^{1.5}+fe^{y/wy}+ge^{-y}$
161	0.7963206646	58	**39793843**	$z=a+bx+cx^{1.5}+dx^{0.5}lnx+ey^{1.5}+fy^{2.5}+ge^{y/wy}+h(lny)^{2}+ie^{-y}$
162	0.7963206646	30	**140846**	$z=a+bx+cx^{1.5}+dx^{0.5}lnx+ey^{1.5}+fy^{2.5}$
163	0.7963206646	39	**1115235**	$z=a+bx+cx^{1.5}+dx^{0.5}lnx+ey^{1.5}+fy^{2.5}+g(lny)^{2}$
164	0.7963206646	28	**149390**	$z=a+bx+cx^{2}+dx^{0.5}lnx+ey^{1.5}+fy^{2.5}$
165	0.7963206646	41	**1115233**	$z=a+bx+cx^{1.5}+dx^{0.5}lnx+ey^{1.5}+fy^{2.5}+ge^{y/wy}$
166	0.7963206646	50	**7286022**	$z=a+bx+cx^{1.5}+dx^{0.5}lnx+ey^{1.5}+fy^{2.5}+ge^{y/wy}+h(lny)^{2}$
167	0.7963206646	39	**1115246**	$z=a+bx+cx^{1.5}+dx^{0.5}lnx+ey^{1.5}+fy^{2.5}+ge^{-y}$

168	0.7963206646	41	**1115296**	$z=a+bx+cx^{1.5}+dx^{0.5}\ln x+ey^{1.5}+f(\ln y)^2+ge^{-y}$
169	0.7963206646	48	**7286056**	$z=a+bx+cx^{1.5}+dx^{0.5}\ln x+ey^{1.5}+fy^{2.5}+g(\ln y)^2+he^{-y}$
170	0.7963206646	60	**39786077**	$z=a+bx+cx^{1.5}+dx^{0.5}\ln x+ey^{1.5}+fy^2\ln y+gy^{2.5}+he^{y/wy}+ie^{-y}$
171	0.7963206646	48	**6774329**	$z=a+bx+cx\ln x+dx^{0.5}+ey^{1.5}+fy^2\ln y+g(\ln y)^2+he^{-y}$
172	0.7963206646	57	**13811914**	$z=a+bx\ln x+cx^{0.5}\ln x+dx^{0.5}+ey^{1.5}+fy^2\ln y+g(\ln y)^2+he^{-y}$
173	0.7963206646	59	**35892349**	$z=a+bx+cx\ln x+dx^{0.5}\ln x+ex^{0.5}+fy^{1.5}+gy^2\ln y+h(\ln y)^2+ie^{-y}$
174	0.7963206646	44	**1115985**	$z=a+bx+cx^{1.5}+dx^{0.5}\ln x+ey^{2.5}+fe^{y/wy}+ge^{-y}$
175	0.7963206646	45	**1024637**	$z=a+bx+cx\ln x+dx^{0.5}\ln x+ex^{0.5}+fy^2\ln y+ge^{-y}$
176	0.7963206646	62	**39787298**	$z=a+bx+cx^{1.5}+dx^{0.5}\ln x+ey^{1.5}+fy^2\ln y+ge^{y/wy}+h(\ln y)^2+ie^{-y}$
177	0.7963206646	56	**35892288**	$z=a+bx+cx\ln x+dx^{0.5}\ln x+ex^{0.5}+fy^{1.5}+gy^2\ln y+hy^{2.5}+i(\ln y)^2$
178	0.7963206646	57	**35892299**	$z=a+bx+cx\ln x+dx^{0.5}\ln x+ex^{0.5}+fy^{1.5}+gy^2\ln y+hy^{2.5}+ie^{-y}$
179	0.7963206646	61	**35892326**	$z=a+bx+cx\ln x+dx^{0.5}\ln x+ex^{0.5}+fy^{1.5}+gy^2\ln y+he^{y/wy}+ie^{-y}$
180	0.7963206646	61	**36225976**	$z=a+bx+cx\ln x+dx^{0.5}\ln x+ey^{1.5}+fy^2\ln y+gy^{2.5}+h(\ln y)^2+ie^{-y}$
181	0.7963206646	24	**219**	$z=a+b/x+c/x^2+d/x^3+e/x^4+f/x^5+g\ln y+h(\ln y)^2+i(\ln y)^3+j(\ln y)^4$
182	0.7963206646	32	**82434**	$z=a+bx^{0.5}\ln x+cx^{0.5}+d\ln x+e(\ln y)^2$
183	0.7963206646	29	**82430**	$z=a+bx^{0.5}\ln x+cx^{0.5}+d\ln x+ey^{2.5}$
184	0.7963206646	33	**82443**	$z=a+bx^{0.5}\ln x+cx^{0.5}+d\ln x+e\ln y/y^2$
185	0.7963206646	32	**82426**	$z=a+bx^{0.5}\ln x+cx^{0.5}+d\ln x+ey\ln y$
186	0.7963206646	32	**82445**	$z=a+bx^{0.5}\ln x+cx^{0.5}+d\ln x+ee^{-y}$
187	0.7963206646	19	**200**	$z=a+b/x+c/x^2+d/x^3+e/x^4+fy+gy^2+hy^3+iy^4+jy^5$
188	0.7963206646	33	**82432**	$z=a+bx^{0.5}\ln x+cx^{0.5}+d\ln x+ee^{y/wy}$
189	0.7963206646	33	**82429**	$z=a+bx^{0.5}\ln x+cx^{0.5}+d\ln x+ey^2\ln y$
190	0.7963206646	20	**189**	$z=a+b/x+c/x^2+d/x^3+e\ln y+f(\ln y)^2+g(\ln y)^3+h(\ln y)^4$
191	0.7963206646	22	**204**	$z=a+b/x+c/x^2+d/x^3+e/x^4+f\ln y+g(\ln y)^2+h(\ln y)^3+i(\ln y)^4$
192	0.7963206646	28	**82427**	$z=a+bx^{0.5}\ln x+cx^{0.5}+d\ln x+ey^{1.5}$
193	0.7963206646	32	**82435**	$z=a+bx^{0.5}\ln x+cx^{0.5}+d\ln x+ey/\ln y$
194	0.7963206646	27	**26835**	$z=a+bx\ln x+cx^2+dx^{2.5}+ey\ln y$
195	0.7963206646	21	**138**	$z=a+b\ln x+c(\ln x)^2+d(\ln x)^3+e(\ln x)^4+f(\ln x)^5+gy+hy^2+iy^3$
196	0.7963206646	24	**205**	$z=a+b/x+c/x^2+d/x^3+e/x^4+f\ln y+g(\ln y)^2+h(\ln y)^3+i(\ln y)^4+j(\ln y)^5$
197	0.7963206646	13	**183**	$z=a+b/x+c/x^2+d/x^3+ey+fy^2+gy^3$
198	0.7963206646	24	**142**	$z=a+b\ln x+c(\ln x)^2+d(\ln x)^3+e(\ln x)^4+f(\ln x)^5+g\ln y+h(\ln y)^2$
199	0.7963206646	17	**66**	$z=a+bx+cx^2+dx^3+ex^4+fx^5+g\ln y$
200	0.7963206646	19	**67**	$z=a+bx+cx^2+dx^3+ex^4+fx^5+g\ln y+h(\ln y)^2$
201	0.7963206646	20	**203**	$z=a+b/x+c/x^2+d/x^3+e/x^4+f\ln y+g(\ln y)^2+h(\ln y)^3$
202	0.7963206646	20	**112**	$z=a+b\ln x+c(\ln x)^2+d(\ln x)^3+e\ln y+f(\ln y)^2$
203	0.7963206646	22	**127**	$z=a+b\ln x+c(\ln x)^2+d(\ln x)^3+e(\ln x)^4+f\ln y+g(\ln y)^2$
204	0.7963206646	20	**217**	$z=a+b/x+c/x^2+d/x^3+e/x^4+f/x^5+g\ln y+h(\ln y)^2$
205	0.7963206646	24	**114**	$z=a+b\ln x+c(\ln x)^2+d(\ln x)^3+e\ln y+f(\ln y)^2+g(\ln y)^3+h(\ln y)^4$
206	0.7963206646	15	**37**	$z=a+bx+cx^2+dx^3+e\ln y+f(\ln y)^2$
207	0.7963206646	17	**185**	$z=a+b/x+c/x^2+d/x^3+ey+fy^2+gy^3+hy^4+iy^5$
208	0.7963206646	18	**202**	$z=a+b/x+c/x^2+d/x^3+e/x^4+f\ln y+g(\ln y)^2$
209	0.7963206646	17	**52**	$z=a+bx+cx^2+dx^3+ex^4+f\ln y+g(\ln y)^2$
210	0.7963206646	15	**198**	$z=a+b/x+c/x^2+d/x^3+e/x^4+fy+gy^2+hy^3$
211	0.7963206646	16	**187**	$z=a+b/x+c/x^2+d/x^3+e\ln y+f(\ln y)^2$
212	0.7963206646	20	**224**	$z=a+b/x+c/x^2+d/x^3+e/x^4+f/x^5+g/y+h/y^2+i/y^3+j/y^4$
213	0.7963206646	16	**201**	$z=a+b/x+c/x^2+d/x^3+e/x^4+f\ln y$
214	0.7963206646	15	**72**	$z=a+bx+cx^2+dx^3+ex^4+fx^5+g/y+h/y^2$
215	0.7963206646	16	**194**	$z=a+b/x+c/x^2+d/x^3+e/y+f/y^2+g/y^3+h/y^4$
216	0.7963206646	22	**141**	$z=a+b\ln x+c(\ln x)^2+d(\ln x)^3+e(\ln x)^4+f(\ln x)^5+g\ln y$
217	0.7963206646	17	**213**	$z=a+b/x+c/x^2+d/x^3+e/x^4+f/x^5+gy+hy^2+iy^3$
218	0.7963206646	17	**199**	$z=a+b/x+c/x^2+d/x^3+e/x^4+fy+gy^2+hy^3+iy^4$
219	0.7963206646	28	**130**	$z=a+b\ln x+c(\ln x)^2+d(\ln x)^3+e(\ln x)^4+f\ln y+g(\ln y)^2+h(\ln y)^3+i(\ln y)^4+j(\ln y)^5$
220	0.7963206646	13	**71**	$z=a+bx+cx^2+dx^3+ex^4+fx^5+g/y$
221	0.7963206646	18	**209**	$z=a+b/x+c/x^2+d/x^3+e/x^4+f/y+g/y^2+h/y^3+i/y^4$
222	0.7963206646	9	**41**	$z=a+bx+cx^2+dx^3+e/y$
223	0.7963206646	23	**125**	$z=a+b\ln x+c(\ln x)^2+d(\ln x)^3+e(\ln x)^4+fy+gy^2+hy^3+iy^4+jy^5$
224	0.7963206646	14	**116**	$z=a+b\ln x+c(\ln x)^2+d(\ln x)^3+e/y$
225	0.7963206646	15	**44**	$z=a+bx+cx^2+dx^3+e/y+f/y^2+g/y^3+h/y^4$
226	0.7963206646	18	**146**	$z=a+b\ln x+c(\ln x)^2+d(\ln x)^3+e(\ln x)^4+f(\ln x)^5+g/y$
227	0.7963206646	22	**148**	$z=a+b\ln x+c(\ln x)^2+d(\ln x)^3+e(\ln x)^4+f(\ln x)^5+g/y+h/y^2+i/y^3$
228	0.7963206646	14	**186**	$z=a+b/x+c/x^2+d/x^3+e\ln y$

229	0.7963206646	24	**149**	$z=a+blnx+c(lnx)^2+d(lnx)^3+e(lnx)^4+f(lnx)^5+g/y+h/y^2+i/y^3+j/y^4$
230	0.7963206646	18	**223**	$z=a+b/x+c/x^2+d/x^3+e/x^4+f/x^5+g/y+h/y^2+i/y^3$
231	0.7963206646	12	**33**	$z=a+bx+cx^2+dx^3+ey+fy^2+gy^3$
232	0.7963206646	20	**65**	$z=a+bx+cx^2+dx^3+ex^4+fx^5+gy+hy^2+iy^3+jy^4+ky^5$
233	0.7963206646	13	**43**	$z=a+bx+cx^2+dx^3+e/y+f/y^2+g/y^3$
234	0.7963206646	18	**50**	$z=a+bx+cx^2+dx^3+ex^4+fy+gy^2+hy^3+iy^4+jy^5$
235	0.7963206646	16	**63**	$z=a+bx+cx^2+dx^3+ex^4+fx^5+gy+hy^2+iy^3$
236	0.7963206646	16	**35**	$z=a+bx+cx^2+dx^3+ey+fy^2+gy^3+hy^4+iy^5$
237	0.7963206646	13	**106**	$z=a+blnx+c(lnx)^2+d(lnx)^3+ey$
238	0.7963206646	16	**131**	$z=a+blnx+c(lnx)^2+d(lnx)^3+e(lnx)^4+f/y$
239	0.7963206646	10	**32**	$z=a+bx+cx^2+dx^3+ey+fy^2$
240	0.7963206646	16	**117**	$z=a+blnx+c(lnx)^2+d(lnx)^3+e/y+f/y^2$
241	0.7963206646	15	**107**	$z=a+blnx+c(lnx)^2+d(lnx)^3+ey+fy^2$
242	0.7963206646	11	**42**	$z=a+bx+cx^2+dx^3+e/y+f/y^2$
243	0.7963206646	21	**75**	$z=a+bx+cx^2+dx^3+ex^4+fx^5+g/y+h/y^2+i/y^3+j/y^4+k/y^5$
244	0.7963206646	13	**36**	$z=a+bx+cx^2+dx^3+elny$
245	0.7963206646	19	**60**	$z=a+bx+cx^2+dx^3+ex^4+f/y+g/y^2+h/y^3+i/y^4+j/y^5$
246	0.7963206646	26	**150**	$z=a+blnx+c(lnx)^2+d(lnx)^3+e(lnx)^4+f(lnx)^5+g/y+h/y^2+i/y^3+j/y^4+k/y^5$
247	0.7963206646	8	**31**	$z=a+bx+cx^2+dx^3+ey$
248	0.7963206646	29	**2031**	z=a+by+LOGNORMX(c,d,e)
249	0.7963206646	102	**2078**	z=a+GCUMX(b,c,d)+GCUMY(e,f,g)
250	0.7963206646	55	**2081**	$z=a+by+cy^2+\mathrm{GCUMX}(d,e,f)$
251	0.7963206646	17	**2007**	z=a+by+GAUSSX(c,d,e)
252	0.7963206646	53	**2079**	z=a+by+GCUMX(c,d,e)
253	0.7963206646	27	**2057**	$z=a+by+cy^2+\mathrm{EXTRVALX}(d,e,f)$
254	0.7963206646	54	**2030**	z=a+LOGNORMX(b,c,d)+LOGNORMY(e,f,g)
255	0.7963206646	26	**2114**	z=a+SIGX(b,c,d)+SIGY(e,f,g)
256	0.7963206646	25	**2055**	z=a+by+EXTRVALX(c,d,e)
257	0.7963206646	10	**2019**	z=a+by+LORX(c,d,e)
258	0.7963206646	15	**2115**	z=a+by+SIGX(c,d,e)
259	0.7963206646	25	**2090**	z=a+LORCUMX(b,c,d)+LORCUMY(e,f,g)
260	0.7963206646	21	**2139**	z=a+by+LDRX(c,d,e)
261	0.7963206646	61	**2105**	$z=a+by+cy^2+\mathrm{LNCUMX}(d,e,f)$
262	0.7963206646	18	**2093**	$z=a+by+cy^2+\mathrm{LORCUMX}(d,e,f)$
263	0.7963206646	42	**2067**	z=a+GAUSSX(b,c,d)+LOGNORMY(e,f,g)
264	0.7963206646	12	**2021**	$z=a+by+cy^2+\mathrm{LORX}(d,e,f)$
265	0.7963206646	19	**2009**	$z=a+by+cy^2+\mathrm{GAUSSX}(d,e,f)$
266	0.7963206646	22	**2129**	$z=a+by+cy^2+\mathrm{EXVCUMX}(d,e,f)$
267	0.7963206646	16	**2018**	z=a+LORX(b,c,d)+LORY(e,f,g)
268	0.7963206646	20	**2127**	z=a+by+EXVCUMX(c,d,e)
269	0.7963206646	37	**2126**	z=a+EXVCUMX(b,c,d)+EXVCUMY(e,f,g)
270	0.7963206646	31	**2045**	$z=a+by+cy^2+\mathrm{LOGISTICX}(d,e,f)$
271	0.7963206646	31	**2033**	$z=a+by+cy^2+\mathrm{LOGNORMX}(d,e,f)$
272	0.7963206646	29	**2043**	z=a+by+LOGISTICX(c,d,e)
273	0.7963206646	24	**2089**	z=LORCUMX(a,b,c)+LORCUMY(d,e,f)
274	0.7963206646	17	**2117**	$z=a+by+cy^2+\mathrm{SIGX}(d,e,f)$
275	0.7963206646	59	**2103**	z=a+by+LNCUMX(c,d,e)
276	0.7963206646	23	**2141**	$z=a+by+cy^2+\mathrm{LDRX}(d,e,f)$
277	0.7963206646	14	**2091**	z=a+by+LORCUMX(c,d,e)
278	0.7963206646	53	**2042**	z=a+LOGISTICX(b,c,d)+LOGISTICY(e,f,g)
279	0.7963206646	38	**2138**	z=a+LDRX(b,c,d)+LDRY(e,f,g)
280	0.7963206646	115	**2102**	z=a+LNCUMX(b,c,d)+LNCUMY(e,f,g)
281	0.7963206646	19	**74**	$z=a+bx+cx^2+dx^3+ex^4+fx^5+g/y+h/y^2+i/y^3+j/y^4$
282	0.7963206646	10	**46**	$z=a+bx+cx^2+dx^3+ex^4+fy$
283	0.7963206646	25	**70**	$z=a+bx+cx^2+dx^3+ex^4+fx^5+glny+h(lny)^2+i(lny)^3+j(lny)^4+k(lny)^5$
284	0.7963206646	30	**145**	$z=a+blnx+c(lnx)^2+d(lnx)^3+e(lnx)^4+f(lnx)^5+glny+h(lny)^2+i(lny)^3+j(lny)^4+k(lny)^5$
285	0.7963206646	21	**215**	$z=a+b/x+c/x^2+d/x^3+e/x^4+f/x^5+gy+hy^2+iy^3+jy^4+ky^5$
286	0.7963206646	22	**225**	$z=a+b/x+c/x^2+d/x^3+e/x^4+f/x^5+g/y+h/y^2+i/y^3+j/y^4+k/y^5$
287	0.7963206646	10	**191**	$z=a+b/x+c/x^2+d/x^3+e/y$
288	0.7963206646	18	**111**	$z=a+blnx+c(lnx)^2+d(lnx)^3+elny$

289	0.7963206646	26	**220**	$z=a+b/x+c/x^2+d/x^3+e/x^4+f/x^5+glny+h(lny)^2+i(lny)^3+j(lny)^4+k(lny)^5$
290	0.7963206646	9	**181**	$z=a+b/x+c/x^2+d/x^3+ey$
291	0.7963192947	19	**7595**	$z=a+bx^{2.5}+cx^3+d(lny)^2$
292	0.7963192947	14	**7602**	$z=a+bx^{2.5}+cx^3+d/y$
293	0.7963192947	20	**7596**	$z=a+bx^{2.5}+cx^3+dy/lny$
294	0.7963192947	15	**7597**	$z=a+bx^{2.5}+cx^3+dy^{0.5}$
295	0.7963192947	16	**7588**	$z=a+bx^{2.5}+cx^3+dy^{1.5}$
296	0.7963192947	19	**7587**	$z=a+bx^{2.5}+cx^3+dylny$
297	0.7963192947	21	**7604**	$z=a+bx^{2.5}+cx^3+dlny/y^2$
298	0.7963192947	20	**7590**	$z=a+bx^{2.5}+cx^3+dy^2lny$
299	0.7963192947	17	**7591**	$z=a+bx^{2.5}+cx^3+dy^{2.5}$
300	0.7963192947	21	**7593**	$z=a+bx^{2.5}+cx^3+de^{y/wy}$
301	0.7963015122	114	**2101**	z=LNCUMX(a,b,c)+LNCUMY(d,e,f)
302	0.7962520059	41	**2066**	z=LOGNORMX(a,b,c)+GAUSSY(d,e,f)
303	0.7962514889	53	**2029**	z=LOGNORMX(a,b,c)+LOGNORMY(d,e,f)
304	0.7959345596	37	**2137**	z=LDRX(a,b,c)+LDRY(d,e,f)
305	0.7958911499	36	**2125**	z=EXVCUMX(a,b,c)+EXVCUMY(d,e,f)
306	0.7958671529	18	**97**	$z=a+blnx+c(lnx)^2+dlny+e(lny)^2$
307	0.7958671529	12	**101**	$z=a+blnx+c(lnx)^2+d/y$
308	0.7958671529	11	**91**	$z=a+blnx+c(lnx)^2+dy$
309	0.7958671529	16	**96**	$z=a+blnx+c(lnx)^2+dlny$
310	0.7949470965	17	**151232673**	lnz=a+blnx+cy
311	0.7949470965	24	**151232677**	$lnz=a+blnx+cy^2lny$
312	0.7949470965	23	**151232674**	lnz=a+blnx+cylny
313	0.7949470965	21	**151232678**	$lnz=a+blnx+cy^{2.5}$
314	0.7949470965	19	**151232679**	$lnz=a+blnx+cy^3$
315	0.7949470965	23	**151232682**	$lnz=a+blnx+c(lny)^2$
316	0.7949470965	24	**151232688**	lnz=a+blnx+clny/y
317	0.7949470965	18	**151232676**	$lnz=a+blnx+cy^2$
318	0.7949470965	24	**151232683**	lnz=a+blnx+cy/lny
319	0.7949470965	24	**151232686**	lnz=a+blnx+c/lny
320	0.7949028391	8	**176**	$z=a+b/x+c/x^2+d/y$
321	0.7949028391	7	**166**	$z=a+b/x+c/x^2+dy$
322	0.7949028391	12	**171**	$z=a+b/x+c/x^2+dlny$
323	0.7947246743	30	**2162**	z=a+POWX(b,c)+POWY(d,e)
324	0.7947246743	17	**2163**	z=a+by+POWX(c,d)
325	0.7947246743	19	**2165**	$z=a+by+cy^2$+POWX(d,e)
326	0.7945954863	7	**26**	$z=a+bx+cx^2+d/y$
327	0.7945954863	6	**16**	$z=a+bx+cx^2+dy$
328	0.7945954863	18	**2094**	$z=a+bx+cx^2$+LORCUMY(d,e,f)
329	0.7945954863	55	**2082**	$z=a+bx+cx^2$+GCUMY(d,e,f)
330	0.7945954863	61	**2106**	$z=a+bx+cx^2$+LNCUMY(d,e,f)
331	0.7945954863	15	**2154**	$z=a+bx+cx^2$+EXPY(d,e)
332	0.7945954863	22	**2130**	$z=a+bx+cx^2$+EXVCUMY(d,e,f)
333	0.7945954863	17	**2118**	$z=a+bx+cx^2$+SIGY(d,e,f)
334	0.7945954863	19	**2166**	$z=a+bx+cx^2$+POWY(d,e)
335	0.7945954863	23	**2142**	$z=a+bx+cx^2$+LDRY(d,e,f)
336	0.7945954863	31	**2046**	$z=a+bx+cx^2$+LOGISTICY(d,e,f)
337	0.7945954863	31	**2034**	$z=a+bx+cx^2$+LOGNORMY(d,e,f)
338	0.7945954863	12	**2022**	$z=a+bx+cx^2$+LORY(d,e,f)
339	0.7945954863	11	**21**	$z=a+bx+cx^2+dlny$
340	0.7945810197	21	**2150**	z=a+EXPX(b,c)+EXPY(d,e)
341	0.7945810197	13	**2151**	z=a+by+EXPX(c,d)
342	0.7945810197	15	**2153**	$z=a+by+cy^2$+EXPX(d,e)
343	0.794535417	101	**2077**	z=GCUMX(a,b,c)+GCUMY(d,e,f)
344	0.7942898253	11	**7**	$z=a+bx+clny+d(lny)^2$
345	0.7942898253	11	**3030**	$z=a+bx+c(lny)^2$
346	0.7942898253	6	**3037**	z=a+bx+c/y
347	0.7942898253	11	**3036**	z=a+bx+clny/y
348	0.7942898253	11	**3022**	z=a+bx+cylny
349	0.7942898253	5	**11**	z=a+bx+c/y

350	0.7942898253	7	**3040**	$z=a+bx+c/y^2$
351	0.7942898253	12	**3025**	$z=a+bx+cy^2\ln y$
352	0.7942898253	4	**1**	z=a+bx+cy
353	0.7942898253	9	**3026**	$z=a+bx+cy^{2.5}$
354	0.7942898253	11	**3031**	z=a+bx+cy/lny
355	0.7942898253	5	**3024**	$z=a+bx+cy^2$
356	0.7942898253	4	**3021**	z=a+bx+cy
357	0.7942898253	13	**2152**	z=a+bx+EXPY(c,d)
358	0.7942898253	10	**2020**	z=a+bx+LORY(c,d,e)
359	0.7942898253	4	**2169**	z=a+bx+cy [Robust None, Least Squares]
360	0.7942898253	15	**2116**	z=a+bx+SIGY(c,d,e)
361	0.7942898253	17	**2164**	z=a+bx+POWY(c,d)
362	0.7942898253	59	**2104**	z=a+bx+LNCUMY(c,d,e)
363	0.7942898253	29	**2044**	z=a+bx+LOGISTICY(c,d,e)
364	0.7942898253	21	**2140**	z=a+bx+LDRY(c,d,e)
365	0.7942898253	53	**2080**	z=a+bx+GCUMY(c,d,e)
366	0.7942898253	29	**2032**	z=a+bx+LOGNORMY(c,d,e)
367	0.7942898253	14	**2092**	z=a+bx+LORCUMY(c,d,e)
368	0.7942898253	9	**6**	z=a+bx+clny
369	0.7935009832	41	**2065**	z=GAUSSX(a,b,c)+LOGNORMY(d,e,f)
370	0.7929520942	25	**2113**	z=SIGX(a,b,c)+SIGY(d,e,f)
371	0.7912822255	20	**2145**	z=EXPX(a,b)*EXPY(1,c)
372	0.7907531892	52	**2041**	z=LOGISTICX(a,b,c)+LOGISTICY(d,e,f)
373	0.778198553	29	**2161**	z=POWX(a,b)+POWY(c,d)
374	0.774888121	4	**2170**	z=a+bx+cy [Robust Low, Least Abs Deviation]
375	0.7586920852	10	**86**	z=a+blnx+c/y
376	0.7586920852	9	**76**	z=a+blnx+cy
377	0.7586920852	14	**81**	z=a+blnx+clny
378	0.716700081	29	**2159**	z=POWX(a,b)*POWY(1,b)
379	0.7133502878	16	**302462050**	$z^{-1}=a+b/x^{1.5}+c(\ln y)^2$
380	0.7133502878	9	**302462041**	$z^{-1}=a+b/x^{1.5}+cy$
381	0.7133502878	11	**302462057**	$z^{-1}=a+b/x^{1.5}+c/y$
382	0.7133502878	17	**302462045**	$z^{-1}=a+b/x^{1.5}+cy^2\ln y$
383	0.7133502878	17	**302462051**	$z^{-1}=a+b/x^{1.5}+cy/\ln y$
384	0.7133502878	18	**302462059**	$z^{-1}=a+b/x^{1.5}+c\ln y/y^2$
385	0.7133502878	12	**302462052**	$z^{-1}=a+b/x^{1.5}+cy^{0.5}$
386	0.7133502878	13	**302462058**	$z^{-1}=a+b/x^{1.5}+c/y^{1.5}$
387	0.7133502878	17	**302462056**	$z^{-1}=a+b/x^{1.5}+c\ln y/y$
388	0.7133502878	12	**302462060**	$z^{-1}=a+b/x^{1.5}+c/y^2$
389	0.7117173335	20	**2128**	z=a+bx+EXVCUMY(c,d,e)
390	0.704151248	84	**2072**	z=a+LOGNORMX(b,c,d)+GAUSSY(e,f,g)+LOGNORMX(h,c,d)*GAUSSY(1,f,g)
391	0.7039341624	15	**2017**	z=LORX(a,b,c)+LORY(d,e,f)
392	0.7039341597	42	**2068**	z=a+LOGNORMX(b,c,d)+GAUSSY(e,f,g)
393	0.7039341575	30	**2006**	z=a+GAUSSX(b,c,d)+GAUSSY(e,f,g)
394	0.7039341197	60	**2012**	z=a+GAUSSX(b,c,d)+GAUSSY(e,f,g)+GAUSSX(h,c,d)*GAUSSY(1,f,g)
395	0.7022089814	19	**2010**	$z=a+bx+cx^2$+GAUSSY(d,e,f)
396	0.7019033204	17	**2008**	z=a+bx+GAUSSY(c,d,e)
397	0.6921342462	25	**2111**	z=SIGX(a,b,c)*SIGY(1,d,c)
398	0.6919931087	4	**2171**	z=a+bx+cy [Robust Medium, Lorentzian]
399	0.6905013733	4	**2172**	z=a+bx+cy [Robust High, PearsonVII Limit]
400	0.6819763105	6	**161**	z=a+b/x+c/y
401	0.6819763105	5	**151**	z=a+b/x+cy
402	0.6819763105	10	**156**	z=a+b/x+clny
403	0.6021133275	30	**2160**	z=a+POWX(b,c)*POWY(1,c)
404	0.5464145043	37	**2124**	z=a+EXVCUMX(b,c,d)*EXVCUMY(1,e,d)
405	0.4659663257	43	**2053**	z=EXTRVALX(a,b,c)+EXTRVALY(d,e,f)
406	0.4307848318	20	**1079**	$z=(a+bx+cx^2+d\ln y)/(1+ex+fx^2+gx^3+h\ln y)$
407	0.4307848318	25	**1095**	$z=(a+bx+cx^2+d\ln y+e(\ln y)^2+f(\ln y)^3)/(1+gx+hx^2+ix^3+j\ln y)$
408	0.4307848318	22	**1087**	$z=(a+bx+cx^2+d\ln y+e(\ln y)^2)/(1+fx+gx^2+hx^3+i\ln y)$
409	0.3755243677	44	**2054**	z=a+EXTRVALX(b,c,d)+EXTRVALY(e,f,g)

410	0.3671886054	20	**2147**	z=EXPX(a,b)*EXPY(1,b)
411	0.3152131889	17	**1108**	$z=(a+bx+cx^2+dx^3+ey+fy^2)/(1+gx+hx^2+iy)$
412	0.3152131889	15	**1100**	$z=(a+bx+cx^2+dx^3+ey)/(1+fx+gx^2+hy)$
413	0.3151558282	22	**1112**	$z=(a+bx+cx^2+dx^3+ey+fy^2)/(1+gx+hx^2+ix^3+jy+ky^2)$
414	0.3151558282	20	**1104**	$z=(a+bx+cx^2+dx^3+ey)/(1+fx+gx^2+hx^3+iy+jy^2)$
415	0.250378618	25	**2056**	z=a+bx+EXTRVALY(c,d,e)
416	0.2255595908	29	**2005**	z=GAUSSX(a,b,c)+GAUSSY(d,e,f)
417	0.1463874775	29	**1302**	$z=(a+cx+e\ln y+gx^2+i(\ln y)^2+kx\ln y)/(1+bx+d\ln y+fx^2+h(\ln y)^2+jx\ln y)$
418	0.0937473521	27	**2058**	$z=a+bx+cx^2+$EXTRVALY(d,e,f)
419	0.0737210651	91	**2059**	z=EXTRVALX(a,b,c)+EXTRVALY(d,e,f)+EXTRVALX(g,b,c)*EXTRVALY(1,e,f)
420	0.0732514462	25	**1239**	$z=(a+b\ln x+c(\ln x)^2+d(\ln x)^3+ey+fy^2)/(1+g\ln x+h(\ln x)^2+i(\ln x)^3+jy)$
421	0.0732514462	22	**1231**	$z=(a+b\ln x+c(\ln x)^2+d(\ln x)^3+ey)/(1+f\ln x+g(\ln x)^2+h(\ln x)^3+iy)$
422	0.0597827468	102	**2076**	z=a+GCUMX(b,c,d)*GCUMY(1,e,d)
423	0.0453259328	15	**1055**	$z=(a+bx+cx^2+dy)/(1+ex+fx^2+gx^3+hy)$
424	0.0453259328	20	**1071**	$z=(a+bx+cx^2+dy+ey^2+fy^3)/(1+gx+hx^2+ix^3+jy)$
425	0.0453259328	17	**1063**	$z=(a+bx+cx^2+dy+ey^2)/(1+fx+gx^2+hx^3+iy)$
426	0.0090943748	15	**1015**	$z=(a+bx+cy+dy^2)/(1+ex+fx^2+gx^3+hy)$
427	0.0090943748	13	**1007**	$z=(a+bx+cy)/(1+dx+ex^2+fx^3+gy)$
428	0.0090943748	17	**1023**	$z=(a+bx+cy+dy^2+ey^3)/(1+fx+gx^2+hx^3+iy)$

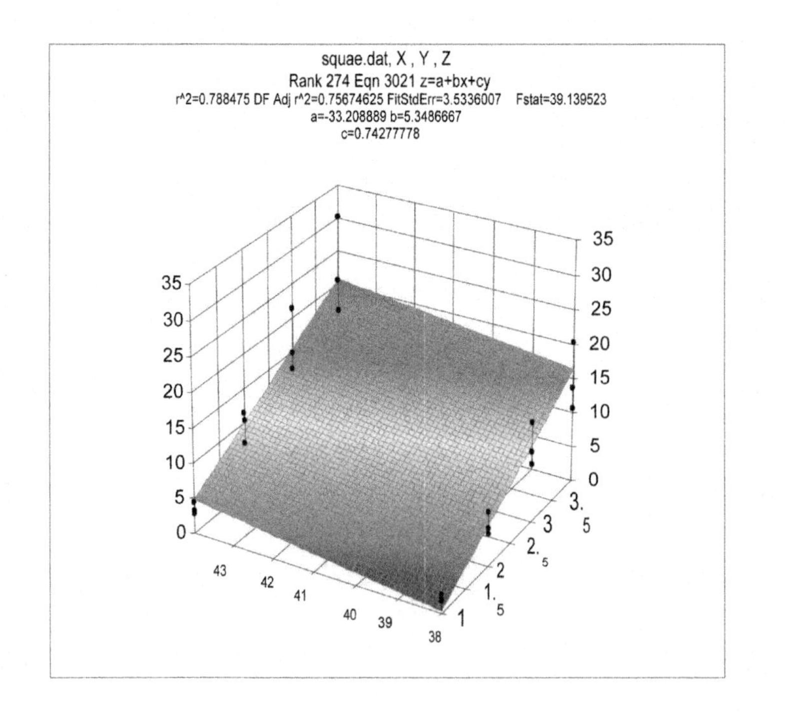

Rank 1 Eqn 314 $z=a+b\ln x+c\ln y+d(\ln x)^2+e(\ln y)^2+f\ln x\ln y+g(\ln x)^3+h(\ln y)^3+i\ln x(\ln y)^2+j(\ln x)^2\ln y$

r^2 Coef Det	DF Adj r^2	Fit Std Err	F-value
0.8133507479	0.6697744001	4.0653260392	6.7785552854

Parm	Value	Std Error	t-value	95.00% Confidence Limits		P>|t|
a	-151.214887					
b	332.5267712					
c	48.06285153					
d	-24.5847885	206.9511565	-0.11879513	-468.450874	419.2812973	0.90713
e	5.257999757					
f	-195.250792					
g	-1.66557336	16.69827402	-0.09974524	-37.4798092	34.1486625	0.92196
h	-1.89372626					
i	28.55353273					
j	9.214837067	54.9063714	0.167828192	-108.547617	126.9772916	0.86912

X at Fn Zmin	Y at Fn Zmin	Fn Zmin
1	38	2.0418201784
X at Fn Zmax	Y at Fn Zmax	Fn Zmax
4	44	22.203303612

Procedure
GaussElim

r^2 Coef Det	DF Adj r^2	Fit Std Err	r^2 Attainable
0.8133507479	0.6697744001	4.0653260392	0.8134711019

Source	Sum of Squares	DF	Mean Square	F Statistic	P>F
Regr	1008.2551	9	112.02834	6.77856	0.00086
Error	231.37626	14	16.526876		
Total	1239.6313	23			
Lack Fit	0.14919461	-2	-0.074597303	-0.00516184	215.64942
Pure Err	231.22707	16	14.451692		

Description: squae.dat, X , Y , Z

X Variable:

Xmin:	1	Xmax:	4	Xrange:	3
Xmean:	2.5	Xstd:	1.1420804814		

Y Variable:

Ymin:	38	Ymax:	44	Yrange:	6
Ymean:	41	Ystd:	3.0645235107		

Z Variable:

Zmin:	1.66	Zmax:	30.11	Zrange:	28.45
Zmean:	10.616666667	Zstd:	7.341458608		

Date	Time	File Source
Apr 6, 2012	1:14:21 PM	d:\documents and settings\samin.samin-e3

squae.dat, X , Y , Z

Rank 1 Eqn 314 z=a+blnx+clny+d(lnx)^2+e(lny)^2+flnxlny+g(lnx)^3+h(lny)^3+ilnx(lny)^2+j(lnx)^2lny

r^2=0.81335075 DF Adj r^2=0.6697744 FitStdErr=4.065326 Fstat=6.7785553

a=-151.21489 b=332.52677 c=48.062852 d=-24.584789 e=5.2579998 f=-195.25079 g=-1.6655734 h=-1.8937263 i=28.553533 j=9.2148371

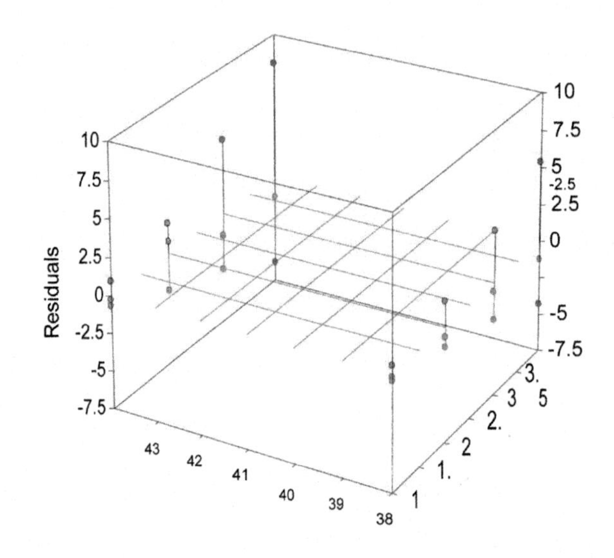

Rank 1 Eqn 314 $z=a+b\ln x+c\ln y+d(\ln x)^2+e(\ln y)^2+f\ln x\ln y+g(\ln x)^3+h(\ln y)^3+i\ln x(\ln y)^2+j(\ln x)^2\ln y$

Precision	Avg Abs Error	Min Abs Error	Max Abs Error
18	2.262284e-16	1.562581e-16	3.262451e-16
17	9.39265e-16	4.739313e-16	1.387819e-15
16	1.007993e-14	2.650741e-16	1.940291e-14
15	8.663042e-14	3.269294e-14	1.424109e-13
14	1.839449e-12	1.26415e-12	2.59963e-12
13	8.343173e-12	5.896905e-12	1.181988e-11
12	1.729305e-10	9.331988e-11	3.17959e-10
11	4.960351e-10	5.454981e-11	1.66304e-09
10	1.653647e-08	8.023023e-09	3.320705e-08
9	5.096794e-08	1.084188e-08	8.638131e-08
8	7.023401e-07	4.650511e-07	1.235789e-06
7	2.563333e-06	3.336806e-07	4.487291e-06
6	7.560473e-05	3.854321e-05	0.0001186854
5	0.0018731176	0.0010825036	0.003278631
4	0.0155558253	0.0032155542	0.0282120188
3	0.2110199337	0.1533096027	0.2820422242
2	0.8969303357	0.3710161753	1.4249226099

Removing	Avg Abs Error	Min Abs Error	Max Abs Error
a	26.918783489	6.8104679326	74.058866126
b	18.302199979	0	36.447382554
c	31.657725795	8.1915262566	85.625935797
d	1.535146277	0	3.1659983614
e	12.81779049	3.3911636438	34.074512244
f	39.777700484	0	77.847706581
g	0.1230291716	0	0.2973469104
h	17.090144701	4.6218758805	44.641581915
i	21.537208329	0	41.411995256
j	2.1298795296	0	4.3166333152

```
/* --------------------------------------------------------------*
                To modify generated output, edit C.TCL
                                                              */

#include <math.h>
#include <stdio.h>

double eqn314(double x, double y);

void main(void)
{
  double  x,y,z;
  char  str[80];
  while(1){
    printf("Enter x:
    "); gets(str);
    if(!*str) break;
    sscanf(str,"%lg",&x)
    ; printf("Enter y:
    "); gets(str);
    if(!*str) break;
    sscanf(str,"%lg",&y)
    ;
    z=eqn314(x,y);
    printf("      z=%.15lg",z);
    }
}
```

```
/*--------------------------------------------------------------*/
double eqn314(double x, double y)
/*--------------------------------------------------------------*

   File Source= d:and settings.samin-
   e39122e4bsquae.dat Date= Apr 6, 2012
   Time= 1:14:23 PM
   Data Set= squae.dat, X , Y ,
   Z X=
   Y=
   Z=
   Eqn#=
   314 Eqn=
z=a+blnx+clny+d(lnx)^2+e(lny)^2+flnxlny+g(lnx)^3+h(lny)^3+ilnx(lny)^2+j
(lnx)^2lny
   r2=0.8133507478785916
   r2adj=0.6697744000928928
   StdErr=4.065326039226049
   Fstat=6.77855528536934
   9 a= -
   151.2148872466162 b=
   332.5267711873423
   c= 48.06285152544785
   d= -24.58478852704998
   e= 5.257999757476177
   f= -195.2507919307165
   g= -1.665573355890993
   h= -1.893726257227051
   i= 28.55353273307606
   j= 9.214837067391996


 *--------------------------------------------------------------*/
{
  double z;
  x=log(x);
  y=log(y);
  z=-151.2148872466162+
    x*(332.5267711873423+x*(-24.58478852704998+
    x*(-1.665573355890993)))+y*(48.06285152544785+
    y*(5.257999757476177+y*(-1.893726257227051)))+
    x*y*(-195.2507919307165+y*(28.55353273307606)+
    x*(9.214837067391996));
  return z;
}
```

1	0.8133507479	33	**314**	$z=a+b\ln x+c\ln y+d(\ln x)^2+e(\ln y)^2+f\ln x\ln y+g(\ln x)^3+h(\ln y)^3+i\ln x(\ln y)^2+j(\ln x)^2\ln y$
2	0.8133507479	29	**317**	$z=a+b\ln x+c/y+d(\ln x)^2+e/y^2+f(\ln x)/y+g(\ln x)^3+h/y^3+i(\ln x)/y^2+j(\ln x)^2/y$
3	0.8133507479	28	**311**	$z=a+b\ln x+cy+d(\ln x)^2+ey^2+fy\ln x+g(\ln x)^3+hy^3+iy^2\ln x+jy(\ln x)^2$
4	0.8133479578	24	**316**	$z=a+bx+c/y+dx^2+e/y^2+fx/y+gx^3+h/y^3+ix/y^2+jx^2/y$
5	0.8133479578	23	**310**	$z=a+bx+cy+dx^2+ey^2+fxy+gx^3+hy^3+ixy^2+jx^2y$
6	0.8133479578	28	**313**	$z=a+bx+c\ln y+dx^2+e(\ln y)^2+fx\ln y+gx^3+h(\ln y)^3+ix(\ln y)^2+jx^2\ln y$
7	0.8131410988	107	**2035**	z=LOGNORMX(a,b,c)+LOGNORMY(d,e,f)+LOGNORMX(g,b,c)*LOGNORMY(1,e,f)
8	0.8131410988	50	**2096**	z=a+LORCUMX(b,c,d)+LORCUMY(e,f,g)+LORCUMX(h,c,d)*LORCUMY(1,f,g)
9	0.8131410988	74	**2132**	z=a+EXVCUMX(b,c,d)+EXVCUMY(e,f,g)+EXVCUMX(h,c,d)*EXVCUMY(1,f,g)
10	0.8131410988	204	**2084**	z=a+GCUMX(b,c,d)+GCUMY(e,f,g)+GCUMX(h,c,d)*GCUMY(1,f,g)
11	0.8131410988	49	**2095**	z=LORCUMX(a,b,c)+LORCUMY(d,e,f)+LORCUMX(g,b,c)*LORCUMY(1,e,f)
12	0.8131410988	229	**2108**	z=a+LNCUMX(b,c,d)+LNCUMY(e,f,g)+LNCUMX(h,c,d)*LNCUMY(1,f,g)
13	0.8131410988	228	**2107**	z=LNCUMX(a,b,c)+LNCUMY(d,e,f)+LNCUMX(g,b,c)*LNCUMY(1,e,f)
14	0.8131410988	84	**2071**	z=a+GAUSSX(b,c,d)+LOGNORMY(e,f,g)+GAUSSX(h,c,d)*LOGNORMY(1,f,g)
15	0.8131410988	31	**2023**	z=LORX(a,b,c)+LORY(d,e,f)+LORX(g,b,c)*LORY(1,e,f)
16	0.8131410988	59	**2011**	z=GAUSSX(a,b,c)+GAUSSY(d,e,f)+GAUSSX(g,b,c)*GAUSSY(1,e,f)
17	0.8131410988	108	**2036**	z=a+LOGNORMX(b,c,d)+LOGNORMY(e,f,g)+LOGNORMX(h,c,d)*LOGNORMY(1,f,g)
18	0.8131410988	83	**2069**	z=GAUSSX(a,b,c)+LOGNORMY(d,e,f)+GAUSSX(g,b,c)*LOGNORMY(1,e,f)
19	0.8131410988	106	**2048**	z=a+LOGISTICX(b,c,d)+LOGISTICY(e,f,g)+LOGISTICX(h,c,d)*LOGISTICY(1,f,g)
20	0.8131410988	52	**2120**	z=a+SIGX(b,c,d)+SIGY(e,f,g)+SIGX(h,c,d)*SIGY(1,f,g)
21	0.8131410988	75	**2143**	z=LDRX(a,b,c)+LDRY(d,e,f)+LDRX(g,b,c)*LDRY(1,e,f)
22	0.8131410988	32	**2024**	z=a+LORX(b,c,d)+LORY(e,f,g)+LORX(h,c,d)*LORY(1,f,g)
23	0.8131346101	29	**315**	$z=a+b/x+c\ln y+d/x^2+e(\ln y)^2+f(\ln y)/x+g/x^3+h(\ln y)^3+i(\ln y)^2/x+j(\ln y)/x^2$
24	0.8131346101	24	**312**	$z=a+b/x+cy+d/x^2+ey^2+fy/x+g/x^3+hy^3+iy^2/x+jy/x^2$
25	0.8131346101	25	**318**	$z=a+b/x+c/y+d/x^2+e/y^2+f/(xy)+g/x^3+h/y^3+i/(xy^2)+j/(x^2y)$
26	0.8131336533	36	**2122**	z=a+EXVCUMX(b,c,d)*EXVCUMY(1,e,f)
27	0.8130804035	102	**2074**	z=a+GCUMX(b,c,d)*GCUMY(1,e,f)
28	0.8130737613	73	**2131**	z=EXVCUMX(a,b,c)+EXVCUMY(d,e,f)+EXVCUMX(g,b,c)*EXVCUMY(1,e,f)
29	0.8129361144	41	**2063**	z=a+GAUSSX(b,c,d)*LOGNORMY(1,e,f)
30	0.8129279563	26	**2110**	z=a+SIGX(b,c,d)*SIGY(1,e,f)
31	0.8129279563	26	**2112**	z=a+SIGX(b,c,d)*SIGY(1,e,d)
32	0.81286469	38	**2134**	z=a+LDRX(b,c,d)*LDRY(1,e,f)
33	0.8128425898	21	**305**	$z=a+b\ln x+c\ln y+d(\ln x)^2+e(\ln y)^2+f\ln x\ln y$
34	0.8128425898	17	**308**	$z=a+b\ln x+c/y+d(\ln x)^2+e/y^2+f(\ln x)/y$
35	0.8128425898	16	**302**	$z=a+b\ln x+cy+d(\ln x)^2+ey^2+fy\ln x$
36	0.8128244467	36	**2121**	z=EXVCUMX(a,b,c)*EXVCUMY(1,d,e)
37	0.8128244467	37	**2123**	z=EXVCUMX(a,b,c)*EXVCUMY(1,d,c)
38	0.8128237708	83	**2070**	z=LOGNORMX(a,b,c)+GAUSSY(d,e,f)+LOGNORMX(g,b,c)*GAUSSY(1,e,f)
39	0.8127730502	37	**2133**	z=LDRX(a,b,c)*LDRY(1,d,e)
40	0.8127214275	45	**2026**	z=a+LOGNORMX(b,c,d)*LOGNORMY(1,e,f)
41	0.8126821096	115	**2098**	z=a+LNCUMX(b,c,d)*LNCUMY(1,e,f)
42	0.8126355824	44	**2025**	z=LOGNORMX(a,b,c)*LOGNORMY(1,d,e)
43	0.8126191108	114	**2097**	z=LNCUMX(a,b,c)*LNCUMY(1,d,e)
44	0.8125634921	203	**2083**	z=GCUMX(a,b,c)+GCUMY(d,e,f)+GCUMX(g,b,c)*GCUMY(1,e,f)

45	0.8125509774	60	**2168**	z=a+POWX(b,c)+POWY(d,e)+POWX(f,c)*POWY(1,e)
46	0.8125487847	11	**301**	$z=a+bx+cy+dx^2+ey^2+fxy$
47	0.8125487847	12	**307**	$z=a+bx+c/y+dx^2+e/y^2+fx/y$
48	0.8125487847	16	**304**	$z=a+bx+c\ln y+dx^2+e(\ln y)^2+fx\ln y$
49	0.8125067515	42	**2156**	z=a+EXPX(b,c)+EXPY(d,e)+EXPX(f,c)*EXPY(1,e)
50	0.8124537861	44	**2040**	z=a+LOGISTICX(b,c,d)*LOGISTICY(1,e,d)
51	0.8124537861	44	**2038**	z=a+LOGISTICX(b,c,d)*LOGISTICY(1,e,f)
52	0.8124506972	44	**2027**	z=LOGNORMX(a,b,c)*LOGNORMY(1,d,c)
53	0.812420562	101	**2073**	z=GCUMX(a,b,c)*GCUMY(1,d,e)
54	0.8124181349	30	**2158**	z=a+POWX(b,c)*POWY(1,d)
55	0.8121106923	115	**2100**	z=a+LNCUMX(b,c,d)*LNCUMY(1,e,d)
56	0.8119040733	40	**2061**	z=GAUSSX(a,b,c)*LOGNORMY(1,d,e)
57	0.8118838931	51	**2119**	z=SIGX(a,b,c)+SIGY(d,e,f)+SIGX(g,b,c)*SIGY(1,e,f)
58	0.8118222972	25	**2109**	z=SIGX(a,b,c)*SIGY(1,d,e)
59	0.8115373463	45	**2028**	z=a+LOGNORMX(b,c,d)*LOGNORMY(1,e,d)
60	0.8110316288	25	**2088**	z=a+LORCUMX(b,c,d)*LORCUMY(1,e,d)
61	0.8110316288	25	**2086**	z=a+LORCUMX(b,c,d)*LORCUMY(1,e,f)
62	0.8106755929	13	**2014**	z=a+LORX(b,c,d)*LORY(1,e,f)
63	0.8106755929	13	**2016**	z=a+LORX(b,c,d)*LORY(1,e,d)
64	0.8104781262	59	**2167**	z=POWX(a,b)+POWY(c,d)+POWX(e,b)*POWY(1,d)
65	0.8103167272	105	**2047**	z=LOGISTICX(a,b,c)+LOGISTICY(d,e,f)+LOGISTICX(g,b,c)*LOGISTICY(1,e,f)
66	0.8103137901	43	**2039**	z=LOGISTICX(a,b,c)*LOGISTICY(1,d,c)
67	0.8103137901	43	**2037**	z=LOGISTICX(a,b,c)*LOGISTICY(1,d,e)
68	0.8100244463	29	**2157**	z=POWX(a,b)*POWY(1,c)
69	0.8095321082	114	**2099**	z=LNCUMX(a,b,c)*LNCUMY(1,d,c)
70	0.8093958821	38	**2136**	z=a+LDRX(b,c,d)*LDRY(1,e,d)
71	0.8093390881	17	**306**	$z=a+b/x+c\ln y+d/x^2+e(\ln y)^2+f(\ln y)/x$
72	0.8093390881	12	**303**	$z=a+b/x+cy+d/x^2+ey^2+fy/x$
73	0.8093390881	13	**309**	$z=a+b/x+c/y+d/x^2+e/y^2+f/(xy)$
74	0.8059386193	37	**2135**	z=LDRX(a,b,c)*LDRY(1,d,c)
75	0.8046607967	41	**2155**	z=EXPX(a,b)+EXPY(c,d)+EXPX(e,b)*EXPY(1,d)
76	0.8041146621	12	**2013**	z=LORX(a,b,c)*LORY(1,d,e)
77	0.8041146621	12	**2015**	z=LORX(a,b,c)*LORY(1,d,c)
78	0.8034679146	24	**2085**	z=LORCUMX(a,b,c)*LORCUMY(1,d,e)
79	0.799834644	21	**2146**	z=a+EXPX(b,c)*EXPY(1,d)
80	0.7891984552	51	**7285441**	$z=a+bx+cx^{1.5}+dx^{0.5}\ln x+ey^{1.5}+fy^2\ln y+g(\ln y)^2+he^{-y}$
81	0.7891984552	35	**149438**	$z=a+bx+cx^2+dx^{0.5}\ln x+ey^2\ln y+fe^{-y}$
82	0.7891984552	40	**1115216**	$z=a+bx+cx^{1.5}+dx^{0.5}\ln x+ey^{1.5}+fy^2\ln y+gy^{2.5}$
83	0.7891984552	32	**149392**	$z=a+bx+cx^2+dx^{0.5}\ln x+ey^{1.5}+fe^{y/wy}$
84	0.7891984552	49	**7285380**	$z=a+bx+cx^{1.5}+dx^{0.5}\ln x+ey^{1.5}+fy^2\ln y+gy^{2.5}+h(\ln y)^2$
85	0.7891984552	37	**140894**	$z=a+bx+cx^{1.5}+dx^{0.5}\ln x+ey^2\ln y+fe^{-y}$
86	0.7891984552	38	**140936**	$z=a+bx+cx^{1.5}+dx^{0.5}\ln x+ee^{y/wy}+fe^{-y}$
87	0.7891984552	34	**140848**	$z=a+bx+cx^{1.5}+dx^{0.5}\ln x+ey^{1.5}+fe^{y/wy}$
88	0.7891984552	36	**149480**	$z=a+bx+cx^2+dx^{0.5}\ln x+ee^{y/wy}+fe^{-y}$
89	0.7891984552	31	**149405**	$z=a+bx+cx^2+dx^{0.5}\ln x+ey^{1.5}+fe^{-y}$
90	0.7891984552	43	**1115262**	$z=a+bx+cx^{1.5}+dx^{0.5}\ln x+ey^{1.5}+fe^{y/wy}+g(\ln y)^2$
91	0.7891984552	52	**7287185**	$z=a+bx+cx^{1.5}+dx^{0.5}\ln x+ey^{1.5}+fe^{y/wy}+g(\ln y)^2+he^{-y}$
92	0.7891984552	40	**1024604**	$z=a+bx+cx\ln x+dx^{0.5}\ln x+ex^{0.5}+fy^{1.5}+ge^{-y}$
93	0.7891984552	51	**7285378**	$z=a+bx+cx^{1.5}+dx^{0.5}\ln x+ey^{1.5}+fy^2\ln y+gy^{2.5}+he^{y/wy}$
94	0.7891984552	49	**7285391**	$z=a+bx+cx^{1.5}+dx^{0.5}\ln x+ey^{1.5}+fy^2\ln y+gy^{2.5}+he^{-y}$
95	0.7891984552	59	**39786066**	$z=a+bx+cx^{1.5}+dx^{0.5}\ln x+ey^{1.5}+fy^2\ln y+gy^{2.5}+he^{y/wy}+i(\ln y)^2$
96	0.7891984552	58	**39786100**	$z=a+bx+cx^{1.5}+dx^{0.5}\ln x+ey^{1.5}+fy^2\ln y+gy^{2.5}+h(\ln y)^2+ie^{-y}$
97	0.7891984552	33	**140861**	$z=a+bx+cx^{1.5}+dx^{0.5}\ln x+ey^{1.5}+fe^{-y}$
98	0.7891984552	53	**7285418**	$z=a+bx+cx^{1.5}+dx^{0.5}\ln x+ey^{1.5}+fy^2\ln y+ge^{y/wy}+he^{-y}$
99	0.7891984552	43	**1115273**	$z=a+bx+cx^{1.5}+dx^{0.5}\ln x+ey^{1.5}+fe^{y/wy}+ge^{-y}$
100	0.7891984552	58	**39793843**	$z=a+bx+cx^{1.5}+dx^{0.5}\ln x+ey^{1.5}+fy^{2.5}+ge^{y/wy}+h(\ln y)^2+ie^{-y}$
101	0.7891984552	30	**140846**	$z=a+bx+cx^{1.5}+dx^{0.5}\ln x+ey^{1.5}+fy^{2.5}$
102	0.7891984552	39	**1115235**	$z=a+bx+cx^{1.5}+dx^{0.5}\ln x+ey^{1.5}+fy^{2.5}+g(\ln y)^2$
103	0.7891984552	41	**1115233**	$z=a+bx+cx^{1.5}+dx^{0.5}\ln x+ey^{1.5}+fy^{2.5}+ge^{y/wy}$
104	0.7891984552	50	**7286022**	$z=a+bx+cx^{1.5}+dx^{0.5}\ln x+ey^{1.5}+fy^{2.5}+ge^{y/wy}+h(\ln y)^2$
105	0.7891984552	28	**149390**	$z=a+bx+cx^2+dx^{0.5}\ln x+ey^{1.5}+fy^{2.5}$

106	0.7891984552	39	**1115246**	$z=a+bx+cx^{1.5}+dx^{0.5}\ln x+ey^{1.5}+fy^{2.5}+ge^{-y}$
107	0.7891984552	41	**1115296**	$z=a+bx+cx^{1.5}+dx^{0.5}\ln x+ey^{1.5}+f(\ln y)^2+ge^{-y}$
108	0.7891984552	48	**7286056**	$z=a+bx+cx^{1.5}+dx^{0.5}\ln x+ey^{1.5}+fy^{2.5}+g(\ln y)^2+he^{-y}$
109	0.7891984552	60	**39786077**	$z=a+bx+cx^{1.5}+dx^{0.5}\ln x+ey^{1.5}+fy^2\ln y+gy^{2.5}+he^{y/wy}+ie^{-y}$
110	0.7891984552	48	**6774329**	$z=a+bx+cx\ln x+dx^{0.5}+ey^{1.5}+fy^2\ln y+g(\ln y)^2+he^{-y}$
111	0.7891984552	57	**13811914**	$z=a+bx\ln x+cx^{0.5}\ln x+dx^{0.5}+ey^{1.5}+fy^2\ln y+g(\ln y)^2+he^{-y}$
112	0.7891984552	59	**35892349**	$z=a+bx+cx\ln x+dx^{0.5}\ln x+ex^{0.5}+fy^{1.5}+gy^2\ln y+h(\ln y)^2+ie^{-y}$
113	0.7891984552	44	**1115985**	$z=a+bx+cx^{1.5}+dx^{0.5}\ln x+ey^{2.5}+fe^{y/wy}+ge^{-y}$
114	0.7891984552	45	**1024637**	$z=a+bx+cx\ln x+dx^{0.5}\ln x+ex^{0.5}+fy^2\ln y+ge^{-y}$
115	0.7891984552	24	**219**	$z=a+b/x+c/x^2+d/x^3+e/x^4+f/x^5+g\ln y+h(\ln y)^2+i(\ln y)^3+j(\ln y)^4$
116	0.7891984552	62	**39787298**	$z=a+bx+cx^{1.5}+dx^{0.5}\ln x+ey^{1.5}+fy^2\ln y+ge^{y/wy}+h(\ln y)^2+ie^{-y}$
117	0.7891984552	56	**35892288**	$z=a+bx+cx\ln x+dx^{0.5}\ln x+ex^{0.5}+fy^{1.5}+gy^2\ln y+hy^{2.5}+i(\ln y)^2$
118	0.7891984552	57	**35892299**	$z=a+bx+cx\ln x+dx^{0.5}\ln x+ex^{0.5}+fy^{1.5}+gy^2\ln y+hy^{2.5}+ie^{-y}$
119	0.7891984552	20	**189**	$z=a+b/x+c/x^2+d/x^3+e\ln y+f(\ln y)^2+g(\ln y)^3+h(\ln y)^4$
120	0.7891984552	61	**35892326**	$z=a+bx+cx\ln x+dx^{0.5}\ln x+ex^{0.5}+fy^{1.5}+gy^2\ln y+he^{y/wy}+ie^{-y}$
121	0.7891984552	61	**36225976**	$z=a+bx+cx\ln x+dx^{0.5}\ln x+ey^{1.5}+fy^2\ln y+gy^{2.5}+h(\ln y)^2+ie^{-y}$
122	0.7891984552	22	**204**	$z=a+b/x+c/x^2+d/x^3+e/x^4+f\ln y+g(\ln y)^2+h(\ln y)^3+i(\ln y)^4$
123	0.7891984552	32	**82434**	$z=a+bx^{0.5}\ln x+cx^{0.5}+d\ln x+e(\ln y)^2$
124	0.7891984552	29	**82430**	$z=a+bx^{0.5}\ln x+cx^{0.5}+d\ln x+ey^{2.5}$
125	0.7891984552	33	**82443**	$z=a+bx^{0.5}\ln x+cx^{0.5}+d\ln x+e\ln y/y^2$
126	0.7891984552	32	**82426**	$z=a+bx^{0.5}\ln x+cx^{0.5}+d\ln x+ey\ln y$
127	0.7891984552	32	**82445**	$z=a+bx^{0.5}\ln x+cx^{0.5}+d\ln x+ee^{-y}$
128	0.7891984552	33	**82432**	$z=a+bx^{0.5}\ln x+cx^{0.5}+d\ln x+ee^{y/wy}$
129	0.7891984552	33	**82429**	$z=a+bx^{0.5}\ln x+cx^{0.5}+d\ln x+ey^2\ln y$
130	0.7891984552	23	**21524**	$z=a+bx+cx^{0.5}+d\ln x+e(\ln y)^2$
131	0.7891984552	28	**82427**	$z=a+bx^{0.5}\ln x+cx^{0.5}+d\ln x+ey^{1.5}$
132	0.7891984552	29	**31719**	$z=a+bx\ln x+cx^{0.5}+d\ln x+e(\ln y)^2$
133	0.7891984552	24	**205**	$z=a+b/x+c/x^2+d/x^3+e/x^4+f\ln y+g(\ln y)^2+h(\ln y)^3+i(\ln y)^4+j(\ln y)^5$
134	0.7891984552	24	**142**	$z=a+b\ln x+c(\ln x)^2+d(\ln x)^3+e(\ln x)^4+f(\ln x)^5+g\ln y+h(\ln y)^2$
135	0.7891984552	13	**183**	$z=a+b/x+c/x^2+d/x^3+ey+fy^2+gy^3$
136	0.7891984552	20	**203**	$z=a+b/x+c/x^2+d/x^3+e/x^4+f\ln y+g(\ln y)^2+h(\ln y)^3$
137	0.7891984552	19	**67**	$z=a+bx+cx^2+dx^3+ex^4+fx^5+g\ln y+h(\ln y)^2$
138	0.7891984552	17	**66**	$z=a+bx+cx^2+dx^3+ex^4+fx^5+g\ln y$
139	0.7891984552	20	**112**	$z=a+b\ln x+c(\ln x)^2+d(\ln x)^3+e\ln y+f(\ln y)^2$
140	0.7891984552	22	**127**	$z=a+b\ln x+c(\ln x)^2+d(\ln x)^3+e(\ln x)^4+f\ln y+g(\ln y)^2$
141	0.7891984552	20	**217**	$z=a+b/x+c/x^2+d/x^3+e/x^4+f/x^5+g\ln y+h(\ln y)^2$
142	0.7891984552	24	**114**	$z=a+b\ln x+c(\ln x)^2+d(\ln x)^3+e\ln y+f(\ln y)^2+g(\ln y)^3+h(\ln y)^4$
143	0.7891984552	15	**37**	$z=a+bx+cx^2+dx^3+e\ln y+f(\ln y)^2$
144	0.7891984552	17	**52**	$z=a+bx+cx^2+dx^3+ex^4+f\ln y+g(\ln y)^2$
145	0.7891984552	18	**202**	$z=a+b/x+c/x^2+d/x^3+e/x^4+f\ln y+g(\ln y)^2$
146	0.7891984552	22	**141**	$z=a+b\ln x+c(\ln x)^2+d(\ln x)^3+e(\ln x)^4+f(\ln x)^5+g\ln y$
147	0.7891984552	16	**187**	$z=a+b/x+c/x^2+d/x^3+e\ln y+f(\ln y)^2$
148	0.7891984552	16	**201**	$z=a+b/x+c/x^2+d/x^3+e/x^4+f\ln y$
149	0.7891984552	17	**199**	$z=a+b/x+c/x^2+d/x^3+e/x^4+fy+gy^2+hy^3+iy^4$
150	0.7891984552	20	**224**	$z=a+b/x+c/x^2+d/x^3+e/x^4+f/x^5+g/y+h/y^2+i/y^3+j/y^4$
151	0.7891984552	16	**194**	$z=a+b/x+c/x^2+d/x^3+e/y+f/y^2+g/y^3+h/y^4$
152	0.7891984552	15	**198**	$z=a+b/x+c/x^2+d/x^3+e/x^4+fy+gy^2+hy^3$
153	0.7891984552	23	**125**	$z=a+b\ln x+c(\ln x)^2+d(\ln x)^3+e(\ln x)^4+fy+gy^2+hy^3+iy^4+jy^5$
154	0.7891984552	17	**213**	$z=a+b/x+c/x^2+d/x^3+e/x^4+f/x^5+gy+hy^2+iy^3$
155	0.7891984552	18	**209**	$z=a+b/x+c/x^2+d/x^3+e/x^4+f/y+g/y^2+h/y^3+i/y^4$
156	0.7891984552	18	**146**	$z=a+b\ln x+c(\ln x)^2+d(\ln x)^3+e(\ln x)^4+f(\ln x)^5+g/y$
157	0.7891984552	9	**41**	$z=a+bx+cx^2+dx^3+e/y$
158	0.7891984552	14	**116**	$z=a+b\ln x+c(\ln x)^2+d(\ln x)^3+e/y$
159	0.7891984552	15	**44**	$z=a+bx+cx^2+dx^3+e/y+f/y^2+g/y^3+h/y^4$
160	0.7891984552	16	**131**	$z=a+b\ln x+c(\ln x)^2+d(\ln x)^3+e(\ln x)^4+f/y$
161	0.7891984552	13	**43**	$z=a+bx+cx^2+dx^3+e/y+f/y^2+g/y^3$
162	0.7891984552	16	**35**	$z=a+bx+cx^2+dx^3+ey+fy^2+gy^3+hy^4+iy^5$
163	0.7891984552	18	**50**	$z=a+bx+cx^2+dx^3+ex^4+fy+gy^2+hy^3+iy^4+jy^5$
164	0.7891984552	20	**65**	$z=a+bx+cx^2+dx^3+ex^4+fx^5+gy+hy^2+iy^3+jy^4+ky^5$
165	0.7891984552	13	**71**	$z=a+bx+cx^2+dx^3+ex^4+fx^5+g/y$
166	0.7891984552	17	**108**	$z=a+b\ln x+c(\ln x)^2+d(\ln x)^3+ey+fy^2+gy^3$
167	0.7891984552	10	**32**	$z=a+bx+cx^2+dx^3+ey+fy^2$

168	0.7891984552	15	**107**	$z=a+b\ln x+c(\ln x)^2+d(\ln x)^3+ey+fy^2$
169	0.7891984552	21	**68**	$z=a+bx+cx^2+dx^3+ex^4+fx^5+g\ln y+h(\ln y)^2+i(\ln y)^3$
170	0.7891984552	18	**223**	$z=a+b/x+c/x^2+d/x^3+e/x^4+f/x^5+g/y+h/y^2+i/y^3$
171	0.7891984552	21	**110**	$z=a+b\ln x+c(\ln x)^2+d(\ln x)^3+ey+fy^2+gy^3+hy^4+iy^5$
172	0.7891984552	8	**31**	$z=a+bx+cx^2+dx^3+ey$
173	0.7891984552	16	**63**	$z=a+bx+cx^2+dx^3+ex^4+fx^5+gy+hy^2+iy^3$
174	0.7891984552	19	**123**	$z=a+b\ln x+c(\ln x)^2+d(\ln x)^3+e(\ln x)^4+fy+gy^2+hy^3$
175	0.7891984552	14	**186**	$z=a+b/x+c/x^2+d/x^3+e\ln y$
176	0.7891984552	13	**106**	$z=a+b\ln x+c(\ln x)^2+d(\ln x)^3+ey$
177	0.7891984552	10	**46**	$z=a+bx+cx^2+dx^3+ex^4+fy$
178	0.7891984552	22	**148**	$z=a+b\ln x+c(\ln x)^2+d(\ln x)^3+e(\ln x)^4+f(\ln x)^5+g/y+h/y^2+i/y^3$
179	0.7891984552	24	**149**	$z=a+b\ln x+c(\ln x)^2+d(\ln x)^3+e(\ln x)^4+f(\ln x)^5+g/y+h/y^2+i/y^3+j/y^4$
180	0.7891984552	11	**42**	$z=a+bx+cx^2+dx^3+e/y+f/y^2$
181	0.7891984552	25	**140**	$z=a+b\ln x+c(\ln x)^2+d(\ln x)^3+e(\ln x)^4+f(\ln x)^5+gy+hy^2+iy^3+jy^4+ky^5$
182	0.7891984552	26	**150**	$z=a+b\ln x+c(\ln x)^2+d(\ln x)^3+e(\ln x)^4+f(\ln x)^5+g/y+h/y^2+i/y^3+j/y^4+k/y^5$
183	0.7891984552	19	**60**	$z=a+bx+cx^2+dx^3+ex^4+f/y+g/y^2+h/y^3+i/y^4+j/y^5$
184	0.7891984552	21	**75**	$z=a+bx+cx^2+dx^3+ex^4+fx^5+g/y+h/y^2+i/y^3+j/y^4+k/y^5$
185	0.7891984552	16	**117**	$z=a+b\ln x+c(\ln x)^2+d(\ln x)^3+e/y+f/y^2$
186	0.7891984552	24	**135**	$z=a+b\ln x+c(\ln x)^2+d(\ln x)^3+e(\ln x)^4+f/y+g/y^2+h/y^3+i/y^4+j/y^5$
187	0.7891984552	18	**64**	$z=a+bx+cx^2+dx^3+ex^4+fx^5+gy+hy^2+iy^3+jy^4$
188	0.7891984552	22	**2129**	$z=a+by+cy^2$+EXVCUMX(d,e,f)
189	0.7891984552	12	**2021**	$z=a+by+cy^2$+LORX(d,e,f)
190	0.7891984552	21	**2139**	z=a+by+LDRX(c,d,e)
191	0.7891984552	102	**2078**	z=a+GCUMX(b,c,d)+GCUMY(e,f,g)
192	0.7891984552	115	**2102**	z=a+LNCUMX(b,c,d)+LNCUMY(e,f,g)
193	0.7891984552	30	**2006**	z=a+GAUSSX(b,c,d)+GAUSSY(e,f,g)
194	0.7891984552	26	**2114**	z=a+SIGX(b,c,d)+SIGY(e,f,g)
195	0.7891984552	16	**2018**	z=a+LORX(b,c,d)+LORY(e,f,g)
196	0.7891984552	53	**2042**	z=a+LOGISTICX(b,c,d)+LOGISTICY(e,f,g)
197	0.7891984552	54	**2030**	z=a+LOGNORMX(b,c,d)+LOGNORMY(e,f,g)
198	0.7891984552	31	**2033**	$z=a+by+cy^2$+LOGNORMX(d,e,f)
199	0.7891984552	10	**2019**	z=a+by+LORX(c,d,e)
200	0.7891984552	29	**2031**	z=a+by+LOGNORMX(c,d,e)
201	0.7891984552	29	**2043**	z=a+by+LOGISTICX(c,d,e)
202	0.7891984552	61	**2105**	$z=a+by+cy^2$+LNCUMX(d,e,f)
203	0.7891984552	15	**2115**	z=a+by+SIGX(c,d,e)
204	0.7891984552	23	**2141**	$z=a+by+cy^2$+LDRX(d,e,f)
205	0.7891984552	59	**2103**	z=a+by+LNCUMX(c,d,e)
206	0.7891984552	20	**2127**	z=a+by+EXVCUMX(c,d,e)
207	0.7891984552	38	**2138**	z=a+LDRX(b,c,d)+LDRY(e,f,g)
208	0.7891984552	17	**2117**	$z=a+by+cy^2$+SIGX(d,e,f)
209	0.7891984552	37	**2126**	z=a+EXVCUMX(b,c,d)+EXVCUMY(e,f,g)
210	0.7891984552	18	**2093**	$z=a+by+cy^2$+LORCUMX(d,e,f)
211	0.7891984552	27	**2057**	$z=a+by+cy^2$+EXTRVALX(d,e,f)
212	0.7891984552	53	**2079**	z=a+by+GCUMX(c,d,e)
213	0.7891984552	31	**2045**	$z=a+by+cy^2$+LOGISTICX(d,e,f)
214	0.7891984552	25	**2055**	z=a+by+EXTRVALX(c,d,e)
215	0.7891984552	25	**2090**	z=a+LORCUMX(b,c,d)+LORCUMY(e,f,g)
216	0.7891984552	14	**2091**	z=a+by+LORCUMX(c,d,e)
217	0.7891984552	55	**2081**	$z=a+by+cy^2$+GCUMX(d,e,f)
218	0.7891984552	24	**2089**	z=LORCUMX(a,b,c)+LORCUMY(d,e,f)
219	0.7891984552	17	**2007**	z=a+by+GAUSSX(c,d,e)
220	0.7891984552	19	**2009**	$z=a+by+cy^2$+GAUSSX(d,e,f)
221	0.7891984552	23	**69**	$z=a+bx+cx^2+dx^3+ex^4+fx^5+g\ln y+h(\ln y)^2+i(\ln y)^3+j(\ln y)^4$
222	0.7891984552	13	**36**	$z=a+bx+cx^2+dx^3+e\ln y$
223	0.7891984552	30	**145**	$z=a+b\ln x+c(\ln x)^2+d(\ln x)^3+e(\ln x)^4+f(\ln x)^5+g\ln y+h(\ln y)^2+i(\ln y)^3+j(\ln y)^4+k(\ln y)^5$
224	0.7891984552	25	**70**	$z=a+bx+cx^2+dx^3+ex^4+fx^5+g\ln y+h(\ln y)^2+i(\ln y)^3+j(\ln y)^4+k(\ln y)^5$
225	0.7891984552	21	**215**	$z=a+b/x+c/x^2+d/x^3+e/x^4+f/x^5+gy+hy^2+iy^3+jy^4+ky^5$
226	0.7891984552	22	**225**	$z=a+b/x+c/x^2+d/x^3+e/x^4+f/x^5+g/y+h/y^2+i/y^3+j/y^4+k/y^5$

227	0.7891984552	26	**220**	$z=a+b/x+c/x^2+d/x^3+e/x^4+f/x^5+g\ln y+h(\ln y)^2+i(\ln y)^3+j(\ln y)^4+k(\ln y)^5$
228	0.7891984552	18	**111**	$z=a+b\ln x+c(\ln x)^2+d(\ln x)^3+e\ln y$
229	0.7891984552	10	**191**	$z=a+b/x+c/x^2+d/x^3+e/y$
230	0.7891984552	9	**181**	$z=a+b/x+c/x^2+d/x^3+ey$
231	0.7891971197	20	**11376**	$z=a+b\ln x+c/x+d(\ln y)^2$
232	0.7891971197	15	**11383**	$z=a+b\ln x+c/x+d/y$
233	0.7891971197	20	**11382**	$z=a+b\ln x+c/x+d\ln y/y$
234	0.7891971197	16	**11386**	$z=a+b\ln x+c/x+d/y^2$
235	0.7891971197	17	**11384**	$z=a+b\ln x+c/x+d/y^{1.5}$
236	0.7891971197	21	**11385**	$z=a+b\ln x+c/x+d\ln y/y^2$
237	0.7891971197	18	**11372**	$z=a+b\ln x+c/x+dy^{2.5}$
238	0.7891971197	20	**11387**	$z=a+b\ln x+c/x+de^{-y}$
239	0.7891971197	20	**11377**	$z=a+b\ln x+c/x+dy/\ln y$
240	0.7891971197	22	**11374**	$z=a+b\ln x+c/x+de^{y/wy}$
241	0.7890658128	18	**97**	$z=a+b\ln x+c(\ln x)^2+d\ln y+e(\ln y)^2$
242	0.7890658128	12	**101**	$z=a+b\ln x+c(\ln x)^2+d/y$
243	0.7890658128	11	**91**	$z=a+b\ln x+c(\ln x)^2+dy$
244	0.7890658128	16	**96**	$z=a+b\ln x+c(\ln x)^2+d\ln y$
245	0.7889406692	114	**2101**	z=LNCUMX(a,b,c)+LNCUMY(d,e,f)
246	0.7887964122	41	**2066**	z=LOGNORMX(a,b,c)+GAUSSY(d,e,f)
247	0.7887961232	53	**2029**	z=LOGNORMX(a,b,c)+LOGNORMY(d,e,f)
248	0.7884899593	17	**2163**	z=a+by+POWX(c,d)
249	0.7884899593	30	**2162**	z=a+POWX(b,c)+POWY(d,e)
250	0.7884899593	19	**2165**	$z=a+by+cy^2$+POWX(d,e)
251	0.7884769396	6	**16**	$z=a+bx+cx^2+dy$
252	0.7884769396	7	**26**	$z=a+bx+cx^2+d/y$
253	0.7884769396	55	**2082**	$z=a+bx+cx^2$+GCUMY(d,e,f)
254	0.7884769396	18	**2094**	$z=a+bx+cx^2$+LORCUMY(d,e,f)
255	0.7884769396	31	**2034**	$z=a+bx+cx^2$+LOGNORMY(d,e,f)
256	0.7884769396	61	**2106**	$z=a+bx+cx^2$+LNCUMY(d,e,f)
257	0.7884769396	22	**2130**	$z=a+bx+cx^2$+EXVCUMY(d,e,f)
258	0.7884769396	23	**2142**	$z=a+bx+cx^2$+LDRY(d,e,f)
259	0.7884769396	15	**2154**	$z=a+bx+cx^2$+EXPY(d,e)
260	0.7884769396	12	**2022**	$z=a+bx+cx^2$+LORY(d,e,f)
261	0.7884769396	19	**2166**	$z=a+bx+cx^2$+POWY(d,e)
262	0.7884769396	17	**2118**	$z=a+bx+cx^2$+SIGY(d,e,f)
263	0.7884769396	31	**2046**	$z=a+bx+cx^2$+LOGISTICY(d,e,f)
264	0.7884769396	11	**21**	$z=a+bx+cx^2+d\ln y$
265	0.7884768791	15	**2153**	$z=a+by+cy^2$+EXPX(d,e)
266	0.7884768791	13	**2151**	z=a+by+EXPX(c,d)
267	0.7884750036	11	**7**	$z=a+bx+c\ln y+d(\ln y)^2$
268	0.7884750036	11	**3030**	$z=a+bx+c(\ln y)^2$
269	0.7884750036	11	**3036**	$z=a+bx+c\ln y/y$
270	0.7884750036	4	**1**	$z=a+bx+cy$
271	0.7884750036	11	**3022**	$z=a+bx+cy\ln y$
272	0.7884750036	11	**3031**	$z=a+bx+cy/\ln y$
273	0.7884750036	6	**3037**	$z=a+bx+c/y$
274	0.7884750036	4	**3021**	$z=a+bx+cy$
275	0.7884750036	7	**3040**	$z=a+bx+c/y^2$
276	0.7884750036	6	**2**	$z=a+bx+cy+dy^2$
277	0.7884750036	9	**3026**	$z=a+bx+cy^{2.5}$
278	0.7884750036	6	**3027**	$z=a+bx+cy^3$
279	0.7884750036	13	**3028**	$z=a+bx+ce^{y/wy}$
280	0.7884750036	5	**11**	$z=a+bx+c/y$
281	0.7884750036	21	**2140**	z=a+bx+LDRY(c,d,e)
282	0.7884750036	59	**2104**	z=a+bx+LNCUMY(c,d,e)
283	0.7884750036	29	**2032**	z=a+bx+LOGNORMY(c,d,e)
284	0.7884750036	17	**2164**	z=a+bx+POWY(c,d)
285	0.7884750036	14	**2092**	z=a+bx+LORCUMY(c,d,e)
286	0.7884750036	53	**2080**	z=a+bx+GCUMY(c,d,e)
287	0.7884750036	29	**2044**	z=a+bx+LOGISTICY(c,d,e)

288	0.7884750036	13	**2152**	z=a+bx+EXPY(c,d)
289	0.7884750036	4	**2169**	z=a+bx+cy [Robust None, Least Squares]
290	0.7884750036	15	**2116**	z=a+bx+SIGY(c,d,e)
291	0.7884750036	10	**2020**	z=a+bx+LORY(c,d,e)
292	0.7884750036	7	**12**	$z=a+bx+c/y+d/y^2$
293	0.7884750036	9	**6**	z=a+bx+clny
294	0.7884482027	21	**2150**	z=a+EXPX(b,c)+EXPY(d,e)
295	0.788221459	37	**2137**	z=LDRX(a,b,c)+LDRY(d,e,f)
296	0.7880922861	42	**2067**	z=a+GAUSSX(b,c,d)+LOGNORMY(e,f,g)
297	0.7880738426	36	**2125**	z=EXVCUMX(a,b,c)+EXVCUMY(d,e,f)
298	0.7873411608	8	**176**	$z=a+b/x+c/x^2+d/y$
299	0.7860862589	101	**2077**	z=GCUMX(a,b,c)+GCUMY(d,e,f)
300	0.78522095	22	**151232741**	$\ln z=a+b/x^{0.5}+c/\ln y$
301	0.78522095	20	**151232740**	$\ln z=a+b/x^{0.5}+c\ln y$
302	0.78522095	22	**151232743**	$\ln z=a+b/x^{0.5}+c\ln y/y$
303	0.78522095	22	**151232738**	$\ln z=a+b/x^{0.5}+cy/\ln y$
304	0.78522095	21	**151232737**	$\ln z=a+b/x^{0.5}+c(\ln y)^2$
305	0.78522095	18	**151232730**	$\ln z=a+b/x^{0.5}+cy^{1.5}$
306	0.78522095	16	**151232731**	$\ln z=a+b/x^{0.5}+cy^2$
307	0.78522095	24	**151232736**	$\ln z=a+b/x^{0.5}+cy^{0.5}\ln y$
308	0.78522095	21	**151232729**	$\ln z=a+b/x^{0.5}+cy\ln y$
309	0.78522095	22	**151232732**	$\ln z=a+b/x^{0.5}+cy^2\ln y$
310	0.7846783769	41	**2065**	z=GAUSSX(a,b,c)+LOGNORMY(d,e,f)
311	0.784123051	25	**2113**	z=SIGX(a,b,c)+SIGY(d,e,f)
312	0.7813003556	52	**2041**	z=LOGISTICX(a,b,c)+LOGISTICY(d,e,f)
313	0.7812182162	4	**2170**	z=a+bx+cy [Robust Low, Least Abs Deviation]
314	0.7792203299	20	**2145**	z=EXPX(a,b)*EXPY(1,c)
315	0.7773389433	4	**2172**	z=a+bx+cy [Robust High, PearsonVII Limit]
316	0.7697989032	29	**2161**	z=POWX(a,b)+POWY(c,d)
317	0.759268962	10	**86**	z=a+blnx+c/y
318	0.759268962	9	**76**	z=a+blnx+cy
319	0.759268962	14	**81**	z=a+blnx+clny
320	0.7140408152	29	**2159**	z=POWX(a,b)*POWY(1,b)
321	0.7062845329	20	**2128**	z=a+bx+EXVCUMY(c,d,e)
322	0.697875162	12	**302462052**	$z^{-1}=a+b/x^{1.5}+cy^{0.5}$
323	0.697875162	17	**302462054**	$z^{-1}=a+b/x^{1.5}+c/\ln y$
324	0.697875162	16	**302462050**	$z^{-1}=a+b/x^{1.5}+c(\ln y)^2$
325	0.697875162	18	**302462049**	$z^{-1}=a+b/x^{1.5}+cy^{0.5}\ln y$
326	0.697875162	13	**302462058**	$z^{-1}=a+b/x^{1.5}+c/y^{1.5}$
327	0.697875162	9	**302462041**	$z^{-1}=a+b/x^{1.5}+cy$
328	0.697875162	11	**302462044**	$z^{-1}=a+b/x^{1.5}+cy^2$
329	0.697875162	13	**302462043**	$z^{-1}=a+b/x^{1.5}+cy^{1.5}$
330	0.697875162	18	**302462048**	$z^{-1}=a+b/x^{1.5}+ce^{y/wy}$
331	0.697875162	14	**302462046**	$z^{-1}=a+b/x^{1.5}+cy^{2.5}$
332	0.6931190348	84	**2072**	z=a+LOGNORMX(b,c,d)+GAUSSY(e,f,g)+LOGNORMX(h,c,d)*GAUSSY(1,f,g)
333	0.6930640131	42	**2068**	z=a+LOGNORMX(b,c,d)+GAUSSY(e,f,g)
334	0.6930640131	92	**2060**	z=a+EXTRVALX(b,c,d)+EXTRVALY(e,f,g)+EXTRVALX(h,c,d)*EXTRVALY(1,f,g)
335	0.6930639947	60	**2012**	z=a+GAUSSX(b,c,d)+GAUSSY(e,f,g)+GAUSSX(h,c,d)*GAUSSY(1,f,g)
336	0.6930336596	15	**2017**	z=LORX(a,b,c)+LORY(d,e,f)
337	0.6923424975	19	**2010**	$z=a+bx+cx^2$+GAUSSY(d,e,f)
338	0.6923405615	17	**2008**	z=a+bx+GAUSSY(c,d,e)
339	0.6885733615	6	**161**	z=a+b/x+c/y
340	0.6885733615	5	**151**	z=a+b/x+cy
341	0.6885733615	10	**156**	z=a+b/x+clny
342	0.6810663019	25	**2111**	z=SIGX(a,b,c)*SIGY(1,d,c)
343	0.6747097803	4	**2171**	z=a+bx+cy [Robust Medium, Lorentzian]
344	0.5333677003	37	**2124**	z=a+EXVCUMX(b,c,d)*EXVCUMY(1,e,d)
345	0.5237034139	17	**1103**	$z=(a+bx+cx^2+dx^3+ey)/(1+fx+gx^2+hx^3+iy)$
346	0.5237034139	20	**1111**	$z=(a+bx+cx^2+dx^3+ey+fy^2)/(1+gx+hx^2+ix^3+jy)$

347	0.486171477	30	**2160**	z=a+POWX(b,c)*POWY(1,c)
348	0.4503931036	17	**1108**	$z=(a+bx+cx^2+dx^3+ey+fy^2)/(1+gx+hx^2+iy)$
349	0.4503931036	15	**1100**	$z=(a+bx+cx^2+dx^3+ey)/(1+fx+gx^2+hy)$
350	0.4503616134	20	**1104**	$z=(a+bx+cx^2+dx^3+ey)/(1+fx+gx^2+hx^3+iy+jy^2)$
351	0.4503616134	22	**1112**	$z=(a+bx+cx^2+dx^3+ey+fy^2)/(1+gx+hx^2+ix^3+jy+ky^2)$
352	0.4482606283	43	**2053**	z=EXTRVALX(a,b,c)+EXTRVALY(d,e,f)
353	0.3985736842	102	**2076**	z=a+GCUMX(b,c,d)*GCUMY(1,e,d)
354	0.3696432366	44	**2054**	z=a+EXTRVALX(b,c,d)+EXTRVALY(e,f,g)
355	0.3694986463	20	**2147**	z=EXPX(a,b)*EXPY(1,b)
356	0.3520543696	20	**1079**	$z=(a+bx+cx^2+d\ln y)/(1+ex+fx^2+gx^3+h\ln y)$
357	0.3520543696	22	**1087**	$z=(a+bx+cx^2+d\ln y+e(\ln y)^2)/(1+fx+gx^2+hx^3+i\ln y)$
358	0.3520543696	25	**1095**	$z=(a+bx+cx^2+d\ln y+e(\ln y)^2+f(\ln y)^3)/(1+gx+hx^2+ix^3+j\ln y)$
359	0.2177229749	25	**2056**	z=a+bx+EXTRVALY(c,d,e)
360	0.1558671468	29	**2005**	z=GAUSSX(a,b,c)+GAUSSY(d,e,f)
361	0.1096327404	27	**2058**	$z=a+bx+cx^2$+EXTRVALY(d,e,f)
362	0.0167002904	91	**2059**	z=EXTRVALX(a,b,c)+EXTRVALY(d,e,f)+EXTRVALX(g,b,c)*EXTRVALY(1,e,f)

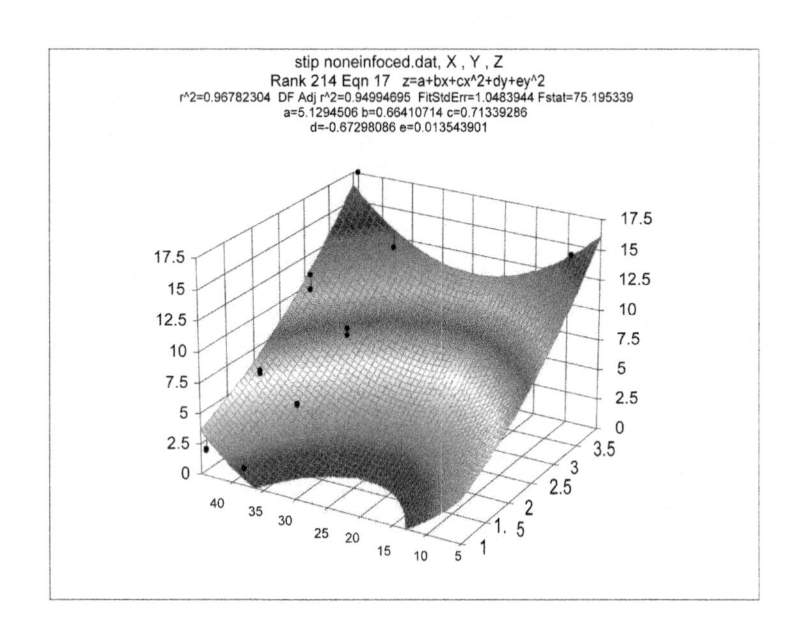

Rank 1 Eqn 316 $z=a+bx+c/y+dx^2+e/y^2+fx/y+gx^3+h/y^3+ix/y^2+jx^2/y$

r^2 Coef Det	DF Adj r^2	Fit Std Err	F-value
0.9970631576	0.9897210515	0.4479272369	188.61208584

Parm	Value	Std Error	t-value	95.00% Confidence Limits		P>|t|
a	173.6062144	2.52196e+07	6.88379e-06	-6.4829e+07	6.48291e+07	0.99999
b	7.793313034					
c	-15862.9582	2.74531e+09	-5.7782e-06	-7.0571e+09	7.05703e+09	1.00000
d	2.111150794	2.128387073	0.991901718	-3.36004237	7.582343958	0.36679
e	425057.7767	8.87552e+10	4.7891e-06	-2.2815e+11	2.28153e+11	1.00000
f	-414.667105					
g	0.08375	0.171055039	0.489608494	-0.35596098	0.523460977	0.64513
h	-2.8044e+06	5.99061e+11	-4.6814e-06	-1.5399e+12	1.53993e+12	1.00000
i	7303.189075					
j	-82.3725397	69.44476695	-1.18615906	-260.885997	96.14091729	0.28885

X at Fn Zmin	Y at Fn Zmin	Fn Zmin
1	39.999133851	1.0009954575
X at Fn Zmax	Y at Fn Zmax	Fn Zmax
4	44	17.467380952

Procedure
GaussElim

r^2 Coef Det	DF Adj r^2	Fit Std Err	r^2 Attainable
0.9970631576	0.9897210515	0.4479272369	0.9970729472

Source	Sum of Squares	DF	Mean Square	F Statistic	P>F
Regr	340.58614	9	37.842904	188.612	0.00001
Error	1.003194	5	0.20063881		
Total	341.58933	14			
Lack Fit	0.0033440476	-1	-0.0033440476	-0.0200673	21.54100
Pure Err	0.99985	6	0.16664167		

Description: stip noneinfoced.dat, X , Y , Z

X Variable:

Xmin:	1	Xmax:	4	Xrange:	3
Xmean:	2.4	Xstd:	1.1212238212		

Y Variable:

Ymin:	9.45	Ymax:	44	Yrange:	34.55
Ymean:	38.896666667	Ystd:	8.6810233652		

Z Variable:

Zmin:	1.38	Zmax:	17.48	Zrange:	16.1
Zmean:	6.9366666667	Zstd:	4.9395584919		

Date	Time	File Source
Apr 6, 2012	1:25:15 PM	d:\documents and settings\samin.samin-e3

stip noneinfoced.dat, X , Y , Z

Rank 1 Eqn 316 z=a+bx+c/y+dx^2+e/y^2+fx/y+gx^3+h/y^3+ix/y^2+jx^2/y

r^2=0.99706316 DF Adj r^2=0.98972105 FitStdErr=0.44792724 Fstat=188.61209

a=173.60621 b=7.793313 c=-15862.958 d=2.1111508 e=425057.78 f=-414.66711

g=0.08375 h=-2804421.8 i=7303.1891 j=-82.37254

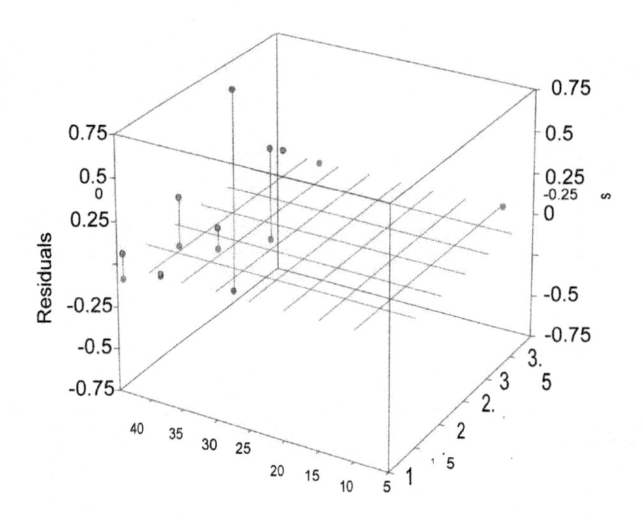

Rank 1 Eqn 316 $z=a+bx+c/y+dx^2+e/y^2+fx/y+gx^3+h/y^3+ix/y^2+jx^2/y$

Precision	Avg Abs Error	Min Abs Error	Max Abs Error
18	3.339869e-16	1.271196e-17	1.659202e-15
17	6.508255e-15	7.237871e-16	2.198958e-14
16	2.917626e-14	2.992077e-15	1.073601e-13
15	3.660868e-13	3.308764e-14	1.552082e-12
14	1.26053e-12	2.359004e-13	4.466113e-12
13	2.067617e-11	4.000748e-12	5.942006e-11
12	4.444552e-10	6.446787e-11	1.259447e-09
11	2.272102e-09	3.895922e-10	6.062798e-09
10	3.89899e-08	7.517281e-09	1.412706e-07
9	2.656112e-07	2.33497e-08	1.204528e-06
8	1.551712e-06	5.553684e-08	7.171896e-06
7	1.715615e-05	2.438789e-06	4.619433e-05
6	0.000222479	1.346606e-07	0.0009994017
5	0.0027441652	0.0002614448	0.0113275895
4	0.0666574909	0.0046747628	0.2891143672
3	0.1360764963	0.029066231	0.4835566411
2	2.3894373816	0.2117428815	9.9931158048

Removing	Avg Abs Error	Min Abs Error	Max Abs Error
a	41.874217006	5.9405222652	156.19371211
b	3.1152772314	0.2666744952	7.0116527592
c	142.07538481	20.639715716	383.36706932
d	1.884295036	0.0722401461	2.7771809142
e	229.42582113	12.569407935	1087.0431482
f	7.3226931345	1.5015062258	24.051460405
g	0.2114329405	0.0028657888	0.4404938963
h	131.90360013	1.88476631	758.94506239
i	9.2243991369	0.8638520948	44.825237859
j	3.655986622	0.2982702983	11.466623813

```
/* ---------------------------------------------------------------*
              To modify generated output, edit C.TCL
                                                                 */

#include <math.h>
#include <stdio.h>

double eqn316(double x, double y);

void main(void)
{
  double  x,y,z;
  char  str[80];
  while(1){
    printf("Enter x:
    "); gets(str);
    if(!*str) break;
    sscanf(str,"%lg",&x)
    ; printf("Enter y:
    "); gets(str);
    if(!*str) break;
    sscanf(str,"%lg",&y)
    ; z=eqn316(x,y);
    printf("      z=%.15lg",z);
    }
}
```

```
/*--------------------------------------------------------------*/
double eqn316(double x, double y)
/*--------------------------------------------------------------*

   File Source= d:and settings.samin-e39122e4bstipnoneinfoced.dat
   Date= Apr 6, 2012
   Time= 1:25:17 PM
   Data Set= stip noneinfoced.dat, X , Y , Z
   X=
   Y=
   Z=
   Eqn#= 316
   Eqn= z=a+bx+c/y+dx^2+e/y^2+fx/y+gx^3+h/y^3+ix/y^2+jx^2/y
   r2=0.9970631575704398
   r2adj=0.9897210514965395
   StdErr=0.4479272368629753
   Fstat=188.6120858417938
   a= 173.6062144272387
   b= 7.793313033827776
   c= -15862.9581952582
   d= 2.111150793652098
   e= 425057.7767247728
   f= -414.6671052820635
   g= 0.0837499999999751
   h= -2804421.782497113
   i= 7303.189075109146
   j= -82.37253968258526
 *--------------------------------------------------------------*/
{
  double z;
  y=1.0/y;
  z=173.6062144272387+
    x*(7.793313033827776+x*(2.111150793652098+
    x*(0.08374999999997510)))+y*(-15862.95819525820+
    y*(425057.7767247728+y*(-2804421.782497113)))+
    x*y*(-414.6671052820635+y*(7303.189075109146)+
    x*(-82.37253968258526));
  return z;
}
```

1	0.9970631576	24	**316**	$z=a+bx+c/y+dx^2+e/y^2+fx/y+gx^3+h/y^3+ix/y^2+jx^2/y$
2	0.9970631576	23	**310**	$z=a+bx+cy+dx^2+ey^2+fxy+gx^3+hy^3+ixy^2+jx^2y$
3	0.9970631576	28	**313**	$z=a+bx+clny+dx^2+e(lny)^2+fxlny+gx^3+h(lny)^3+ix(lny)^2+jx^2lny$
4	0.9969271686	20	**1111**	$z=(a+bx+cx^2+dx^3+ey+fy^2)/(1+gx+hx^2+ix^3+jy)$
5	0.9969006398	25	**1199**	$z=(a+blnx+c(lnx)^2+dy+ey^2+fy^3)/(1+glnx+h(lnx)^2+i(lnx)^3+jy)$
6	0.9969006398	22	**1191**	$z=(a+blnx+c(lnx)^2+dy+ey^2)/(1+flnx+g(lnx)^2+h(lnx)^3+iy)$
7	0.9968710015	29	**317**	$z=a+blnx+c/y+d(lnx)^2+e/y^2+f(lnx)/y+g(lnx)^3+h/y^3+i(lnx)/y^2+j(lnx)^2/y$
8	0.9968710015	28	**311**	$z=a+blnx+cy+d(lnx)^2+ey^2+fylnx+g(lnx)^3+hy^3+iy^2lnx+jy(lnx)^2$
9	0.9968710015	33	**314**	$z=a+blnx+clny+d(lnx)^2+e(lny)^2+flnxlny+g(lnx)^3+h(lny)^3+ilnx(lny)^2+j(lnx)^2lny$
10	0.9968204774	17	**1103**	$z=(a+bx+cx^2+dx^3+ey)/(1+fx+gx^2+hx^3+iy)$
11	0.9968199432	22	**1151**	$z=(a+blnx+cy+dy^2+ey^3)/(1+flnx+g(lnx)^2+h(lnx)^3+iy)$
12	0.9968199432	20	**1143**	$z=(a+blnx+cy+dy^2)/(1+elnx+f(lnx)^2+g(lnx)^3+hy)$
13	0.9968003371	15	**1105**	$z=(a+bx+cx^2+dx^3+ey+fy^2)/(1+gx+hy)$
14	0.9967895532	15	**1065**	$z=(a+bx+cx^2+dy+ey^2+fy^3)/(1+gx+hy)$
15	0.9967895532	13	**1057**	$z=(a+bx+cx^2+dy+ey^2)/(1+fx+gy)$
16	0.9967876333	22	**1236**	$z=(a+blnx+c(lnx)^2+d(lnx)^3+ey+fy^2)/(1+glnx+h(lnx)^2+iy)$
17	0.996760727	25	**1239**	$z=(a+blnx+c(lnx)^2+d(lnx)^3+ey+fy^2)/(1+glnx+h(lnx)^2+i(lnx)^3+jy)$
18	0.9967488771	20	**1064**	$z=(a+bx+cx^2+dy+ey^2)/(1+fx+gx^2+hx^3+iy+jy^2)$
19	0.9967488771	22	**1072**	$z=(a+bx+cx^2+dy+ey^2+fy^3)/(1+gx+hx^2+ix^3+jy+ky^2)$
20	0.9967488683	15	**1060**	$z=(a+bx+cx^2+dy+ey^2)/(1+fx+gx^2+hy)$
21	0.9967488683	17	**1068**	$z=(a+bx+cx^2+dy+ey^2+fy^3)/(1+gx+hx^2+iy)$
22	0.9967242328	18	**1074**	$z=(a+bx+cx^2+dlny)/(1+ex+flny+g(lny)^2)$
23	0.9966779439	20	**1114**	$z=(a+bx+cx^2+dx^3+elny)/(1+fx+glny+h(lny)^2)$
24	0.9966483986	20	**1071**	$z=(a+bx+cx^2+dy+ey^2+fy^3)/(1+gx+hx^2+ix^3+jy)$
25	0.9966483986	17	**1063**	$z=(a+bx+cx^2+dy+ey^2)/(1+fx+gx^2+hx^3+iy)$
26	0.9966407866	27	**1245**	$z=(a+blnx+c(lnx)^2+d(lnx)^3+elny)/(1+flnx+g(lnx)^2+hlny+i(lny)^2)$
27	0.9965969321	22	**1231**	$z=(a+blnx+c(lnx)^2+d(lnx)^3+ey)/(1+flnx+g(lnx)^2+h(lnx)^3+iy)$
28	0.996579065	30	**1255**	$z=(a+blnx+c(lnx)^2+d(lnx)^3+elny+f(lny)^2)/(1+glnx+h(lnx)^2+i(lnx)^3+jlny)$
29	0.9965263772	22	**1441**	Chebyshev X,LnY Rational Order 1/2
30	0.996497413	18	**1029**	$z=(a+bx+clny)/(1+dx+ex^2+flny+g(lny)^2)$
31	0.9964833513	25	**1244**	$z=(a+blnx+c(lnx)^2+d(lnx)^3+elny)/(1+flnx+g(lnx)^2+hlny)$
32	0.9964659001	16	**1026**	$z=(a+bx+clny)/(1+dx+elny+f(lny)^2)$
33	0.9964504914	20	**1039**	$z=(a+bx+clny+d(lny)^2)/(1+ex+fx^2+gx^3+hlny)$
34	0.9964504914	22	**1047**	$z=(a+bx+clny+d(lny)^2+e(lny)^3)/(1+fx+gx^2+hx^3+ilny)$
35	0.996397811	27	**1247**	$z=(a+blnx+c(lnx)^2+d(lnx)^3+elny)/(1+flnx+g(lnx)^2+h(lnx)^3+ilny)$
36	0.9963256114	20	**1024**	$z=(a+bx+cy+dy^2+ey^3)/(1+fx+gx^2+hx^3+iy+jy^2)$
37	0.9963255883	13	**1012**	$z=(a+bx+cy+dy^2)/(1+ex+fx^2+gy)$
38	0.9963255883	15	**1020**	$z=(a+bx+cy+dy^2+ey^3)/(1+fx+gx^2+hy)$
39	0.9963181497	11	**1009**	$z=(a+bx+cy+dy^2)/(1+ex+fy)$
40	0.9963181497	13	**1017**	$z=(a+bx+cy+dy^2+ey^3)/(1+fx+gy)$
41	0.9962967794	25	**1152**	$z=(a+blnx+cy+dy^2+ey^3)/(1+flnx+g(lnx)^2+h(lnx)^3+iy+jy^2)$
42	0.9962965669	20	**1148**	$z=(a+blnx+cy+dy^2+ey^3)/(1+flnx+g(lnx)^2+hy)$
43	0.9962965669	18	**1140**	$z=(a+blnx+cy+dy^2)/(1+elnx+f(lnx)^2+gy)$
44	0.9962785451	20	**1193**	$z=(a+blnx+c(lnx)^2+dy+ey^2+fy^3)/(1+glnx+hy)$
45	0.9962785451	18	**1185**	$z=(a+blnx+c(lnx)^2+dy+ey^2)/(1+flnx+gy)$
46	0.9962387722	25	**1192**	$z=(a+blnx+c(lnx)^2+dy+ey^2)/(1+flnx+g(lnx)^2+h(lnx)^3+iy+jy^2)$
47	0.9962387722	27	**1200**	$z=(a+blnx+c(lnx)^2+dy+ey^2+fy^3)/(1+glnx+h(lnx)^2+i(lnx)^3+jy+ky^2)$
48	0.9961824502	23	**1157**	$z=(a+blnx+clny)/(1+dlnx+e(lnx)^2+flny+g(lny)^2)$
49	0.996172542	18	**1081**	$z=(a+bx+cx^2+dlny+e(lny)^2)/(1+fx+glny)$
50	0.996167673	25	**1048**	$z=(a+bx+clny+d(lny)^2+e(lny)^3)/(1+fx+gx^2+hx^3+ilny+j(lny)^2)$
51	0.9961676631	18	**1036**	$z=(a+bx+clny+d(lny)^2)/(1+ex+fx^2+glny)$
52	0.9961551721	29	**315**	$z=a+b/x+clny+d/x^2+e(lny)^2+f(lny)/x+g/x^3+h(lny)^3+i(lny)^2/x+j(lny)/x^2$
53	0.9961551721	24	**312**	$z=a+b/x+cy+d/x^2+ey^2+fy/x+g/x^3+hy^3+iy^2/x+jy/x^2$
54	0.9961551721	25	**318**	$z=a+b/x+c/y+d/x^2+e/y^2+f/(xy)+g/x^3+h/y^3+i/(xy^2)+j/(x^2y)$

55	0.9960959439	12	**307**	$z=a+bx+c/y+dx^2+e/y^2+fx/y$
56	0.9960959439	30	**401**	Chebyshev X,Y Bivariate Polynomial Order 2
57	0.9960959439	35	**441**	Chebyshev X,LnY Bivariate Polynomial Order 2
58	0.9960959439	11	**301**	$z=a+bx+cy+dx^2+ey^2+fxy$
59	0.9960959439	16	**304**	$z=a+bx+clny+dx^2+e(lny)^2+fxlny$
60	0.9960957954	16	**1033**	$z=(a+bx+clny+d(lny)^2)/(1+ex+flny)$
61	0.9959924115	29	**1302**	$z=(a+cx+elny+gx^2+i(lny)^2+kxlny)/(1+bx+dlny+fx^2+h(lny)^2+jxlny)$
62	0.9958140496	28	**1461**	Chebyshev LnX,LnY Rational Order 1/2
63	0.9956772209	24	**1301**	$z=(a+cx+ey+gx^2+iy^2+kxy)/(1+bx+dy+fx^2+hy^2+jxy)$
64	0.995325912	21	**1154**	$z=(a+blnx+clny)/(1+dlnx+elny+f(lny)^2)$
65	0.9950749231	22	**1112**	$z=(a+bx+cx^2+dx^3+ey+fy^2)/(1+gx+hx^2+ix^3+jy+ky^2)$
66	0.9948519329	16	**1137**	$z=(a+blnx+cy+dy^2)/(1+elnx+fy)$
67	0.9934744484	21	**1161**	$z=(a+blnx+clny+d(lny)^2)/(1+elnx+flny)$
68	0.9934617084	34	**1304**	$z=(a+clnx+elny+g(lnx)^2+i(lny)^2+klnxlny)/(1+blnx+dlny+f(lnx)^2+h(lny)^2+jlnxlny)$
69	0.9929098163	11	**1002**	$z=(a+bx+cy)/(1+dx+ey+fy^2)$
70	0.9918991041	16	**1130**	$z=(a+blnx+cy)/(1+dlnx+ey+fy^2)$
71	0.9917832005	16	**1401**	Chebyshev X,Y Rational Order 1/2
72	0.9889114285	17	**308**	$z=a+blnx+c/y+d(lnx)^2+e/y^2+f(lnx)/y$
73	0.9889114285	16	**302**	$z=a+blnx+cy+d(lnx)^2+ey^2+fylnx$
74	0.9889114285	35	**421**	Chebyshev LnX,Y Bivariate Polynomial Order 2
75	0.9889114285	41	**461**	Chebyshev LnX,LnY Bivariate Polynomial Order 2
76	0.9889114285	21	**305**	$z=a+blnx+clny+d(lnx)^2+e(lny)^2+flnxlny$
77	0.9871682845	52	**2120**	z=a+SIGX(b,c,d)+SIGY(e,f,g)+SIGX(h,c,d)*SIGY(1,f,g)
78	0.9871296375	204	**2084**	z=a+GCUMX(b,c,d)+GCUMY(e,f,g)+GCUMX(h,c,d)*GCUMY(1,f,g)
79	0.9867684964	74	**2132**	z=a+EXVCUMX(b,c,d)+EXVCUMY(e,f,g)+EXVCUMX(h,c,d)*EXVCUMY(1,f,g)
80	0.9866697835	76	**2144**	z=a+LDRX(b,c,d)+LDRY(e,f,g)+LDRX(h,c,d)*LDRY(1,f,g)
81	0.9865277517	73	**2131**	z=EXVCUMX(a,b,c)+EXVCUMY(d,e,f)+EXVCUMX(g,b,c)*EXVCUMY(1,e,f)
82	0.9862565165	203	**2083**	z=GCUMX(a,b,c)+GCUMY(d,e,f)+GCUMX(g,b,c)*GCUMY(1,e,f)
83	0.9856984165	229	**2108**	z=a+LNCUMX(b,c,d)+LNCUMY(e,f,g)+LNCUMX(h,c,d)*LNCUMY(1,f,g)
84	0.9855117572	51	**2119**	z=SIGX(a,b,c)+SIGY(d,e,f)+SIGX(g,b,c)*SIGY(1,e,f)
85	0.9854423855	75	**2143**	z=LDRX(a,b,c)+LDRY(d,e,f)+LDRX(g,b,c)*LDRY(1,e,f)
86	0.9834648129	49	**2095**	z=LORCUMX(a,b,c)+LORCUMY(d,e,f)+LORCUMX(g,b,c)*LORCUMY(1,e,f)
87	0.9822353176	16	**1073**	$z=(a+bx+cx^2+dlny)/(1+ex+flny)$
88	0.9816759722	228	**2107**	z=LNCUMX(a,b,c)+LNCUMY(d,e,f)+LNCUMX(g,b,c)*LNCUMY(1,e,f)
89	0.9809229344	22	**1421**	Chebyshev LnX,Y Rational Order 1/2
90	0.9793885781	29	**1303**	$z=(a+clnx+ey+g(lnx)^2+iy^2+kylnx)/(1+blnx+dy+f(lnx)^2+hy^2+jylnx)$
91	0.9788631329	84	**2071**	z=a+GAUSSX(b,c,d)+LOGNORMY(e,f,g)+GAUSSX(h,c,d)*LOGNORMY(1,f,g)
92	0.9778482545	106	**2048**	z=a+LOGISTICX(b,c,d)+LOGISTICY(e,f,g)+LOGISTICX(h,c,d)*LOGISTICY(1,f,g)
93	0.9768450249	59	**601**	Sigmoid Series Bivariate Order 2
94	0.9764332756	60	**2012**	z=a+GAUSSX(b,c,d)+GAUSSY(e,f,g)+GAUSSX(h,c,d)*GAUSSY(1,f,g)
95	0.973087879	44	**531**	Cosine Series Bivariate Order 2
96	0.972528027	92	**2060**	z=a+EXTRVALX(b,c,d)+EXTRVALY(e,f,g)+EXTRVALX(h,c,d)*EXTRVALY(1,f,g)
97	0.9679919183	13	**309**	$z=a+b/x+c/y+d/x^2+e/y^2+f/(xy)$
98	0.9679919183	12	**303**	$z=a+b/x+cy+d/x^2+ey^2+fy/x$
99	0.9679919183	17	**306**	$z=a+b/x+clny+d/x^2+e(lny)^2+f(lny)/x$
100	0.9679638395	17	**59**	$z=a+bx+cx^2+dx^3+ex^4+f/y+g/y^2+h/y^3+i/y^4$
101	0.9679638395	50	**11541470**	$z=a+bxlnx+cx^2+dx^{2.5}+ey^{0.5}lny+fy/lny+gy^{0.5}+hlny$
102	0.9679638395	62	**78435268**	$z=a+bx^{1.5}+cx^2lnx+dx^{2.5}+e(lnx)^2+fy^{0.5}lny+gy/lny+hy^{0.5}+ilny$

103	0.9679638395	50	**14640046**	$z=a+bx^{1.5}+cx^2+dx^2\ln x+ey^{0.5}\ln y+fy/\ln y+gy^{0.5}+h\ln y$
104	0.9679638395	15	**184**	$z=a+b/x+c/x^2+d/x^3+ey+fy^2+gy^3+hy^4$
105	0.9679638395	47	**14721842**	$z=a+bx^{1.5}+cx^2+dx^{2.5}+ey^{0.5}\ln y+fy/\ln y+gy^{0.5}+h\ln y$
106	0.9679638395	57	**12034976**	$z=a+bx\ln x+cx^2\ln x+dx^{2.5}+ey^{0.5}\ln y+fy/\ln y+gy^{0.5}+h\ln y$
107	0.9679638395	51	**17902214**	$z=a+bx^2+cx^2\ln x+dx^{2.5}+ey^{0.5}\ln y+fy/\ln y+gy^{0.5}+h\ln y$
108	0.9679638395	59	**12312967**	$z=a+bx\ln x+cx^2\ln x+d(\ln x)^2+ey^{0.5}\ln y+fy/\ln y+gy^{0.5}+h\ln y$
109	0.9679638395	49	**14969942**	$z=a+bx^{1.5}+cx^2+d(\ln x)^2+ey^{0.5}\ln y+fy/\ln y+gy^{0.5}+h\ln y$
110	0.9679638395	62	**56332831**	$z=a+bx\ln x+cx^{1.5}+dx^2\ln x+ex^{2.5}+fy^{0.5}\ln y+gy/\ln y+hy^{0.5}+i\ln y$
111	0.9679638395	57	**55718643**	$z=a+bx\ln x+cx^{1.5}+dx^2+ex^2\ln x+f(\ln x)^2+gy/\ln y+hy^{0.5}+i\ln y$
112	0.9679638395	54	**15297599**	$z=a+bx^{1.5}+cx^2\ln x+dx^{2.5}+ey^{0.5}\ln y+fy/\ln y+gy^{0.5}+h\ln y$
113	0.9679638395	50	**18779308**	$z=a+bx^2+cx^{2.5}+d(\ln x)^2+ey^{0.5}\ln y+fy/\ln y+gy^{0.5}+h\ln y$
114	0.9679638395	56	**15575590**	$z=a+bx^{1.5}+cx^2\ln x+d(\ln x)^2+ey^{0.5}\ln y+fy/\ln y+gy^{0.5}+h\ln y$
115	0.9679638395	61	**57031027**	$z=a+bx\ln x+cx^{1.5}+dx^{2.5}+e(\ln x)^2+fy^{0.5}\ln y+gy/\ln y+hy^{0.5}+i\ln y$
116	0.9679638395	65	**56581445**	$z=a+bx\ln x+cx^{1.5}+dx^2\ln x+e(\ln x)^2+fy^{0.5}\ln y+gy/\ln y+hy^{0.5}+i\ln y$
117	0.9679638395	57	**55786229**	$z=a+bx\ln x+cx^{1.5}+dx^2+ex^{2.5}+f(\ln x)^2+gy^{0.5}\ln y+hy/\ln y+iy^{0.5}$
118	0.9679638395	58	**56052064**	$z=a+bx\ln x+cx^{1.5}+dx^2+e(\ln x)^2+fy^{0.5}\ln y+gy/\ln y+hy^{0.5}+i\ln y$
119	0.9679638395	17	**185**	$z=a+b/x+c/x^2+d/x^3+ey+fy^2+gy^3+hy^4+iy^5$
120	0.9679638395	22	**148**	$z=a+b\ln x+c(\ln x)^2+d(\ln x)^3+e(\ln x)^4+f(\ln x)^5+g/y+h/y^2+i/y^3$
121	0.9679638395	60	**56288306**	$z=a+bx\ln x+cx^{1.5}+dx^2\ln x+ex^{2.5}+f(\ln x)^2+g(\ln y)^2+hy^{0.5}+i\ln y$
122	0.9679638395	46	**1730912**	$z=a+bx\ln x+cx^{1.5}+dx^2+e(\ln x)^2+fy^{0.5}\ln y+gy/\ln y$
123	0.9679638395	42	**2388360**	$z=a+bx^{1.5}+cx^2+dx^2\ln x+e(\ln x)^2+fy^{0.5}\ln y+gy^{0.5}$
124	0.9679638395	40	**2388371**	$z=a+bx^{1.5}+cx^2+dx^2\ln x+e(\ln x)^2+f(\ln y)^2+gy^{0.5}$
125	0.9679638395	49	**1747331**	$z=a+bx\ln x+cx^{1.5}+dx^{2.5}+e(\ln x)^2+fy^{0.5}\ln y+gy/\ln y$
126	0.9679638395	62	**55686199**	$z=a+bx\ln x+cx^{1.5}+dx^2+ex^2\ln x+fx^{2.5}+g(\ln x)^2+hy^{0.5}\ln y+iy/\ln y$
127	0.9679638395	53	**1739490**	$z=a+bx\ln x+cx^{1.5}+dx^2\ln x+e(\ln x)^2+fy^{0.5}\ln y+gy/\ln y$
128	0.9679638395	42	**1827513**	$z=a+bx\ln x+cx^2+dx^2\ln x+e(\ln y)^2+fy^{0.5}+g\ln y$
129	0.9679638395	58	**55718523**	$z=a+bx\ln x+cx^{1.5}+dx^2+ex^2\ln x+f(\ln x)^2+g(\ln y)^2+hy/\ln y+iy^{0.5}$
130	0.9679638395	34	**239717**	$z=a+bx\ln x+cx^2+dx^{2.5}+ey^{0.5}\ln y+fy^{0.5}$
131	0.9679638395	39	**3086018**	$z=a+bx^2+cx^{2.5}+d(\ln x)^2+e(\ln y)^2+fy^{0.5}+g\ln y$
132	0.9679638395	41	**1730913**	$z=a+bx\ln x+cx^{1.5}+dx^2+e(\ln x)^2+fy^{0.5}\ln y+gy^{0.5}$
133	0.9679638395	42	**1836203**	$z=a+bx\ln x+cx^2+dx^{2.5}+ey^3+fy^{0.5}\ln y+g(\ln y)^2$
134	0.9679638395	38	**1736201**	$z=a+bx\ln x+cx^{1.5}+dx^2+e(\ln y)^2+fy^{0.5}+g\ln y$
135	0.9679638395	35	**239023**	$z=a+bx\ln x+cx^2+dx^2\ln x+e(\ln y)^2+fy^{0.5}$
136	0.9679638395	42	**239011**	$z=a+bx\ln x+cx^2+dx^2\ln x+ey^{0.5}\ln y+fy/\ln y$
137	0.9679638395	24	**137790**	$z=a+bx+cx^{1.5}+dx^2+e(\ln y)^2+fy^{0.5}$
138	0.9679638395	31	**414604**	$z=a+bx^2+cx^{2.5}+d(\ln x)^2+e(\ln y)^2+fy^{0.5}$
139	0.9679638395	27	**128651**	$z=a+bx+cx\ln x+dx^2+e(\ln y)^2+fy^{0.5}$
140	0.9679638395	35	**405354**	$z=a+bx^2+cx^2\ln x+d(\ln x)^2+e(\ln y)^2+fy^{0.5}$
141	0.9679638395	36	**127936**	$z=a+bx+cx\ln x+dx^{1.5}+ey^{0.5}\ln y+fy/\ln y$
142	0.9679638395	32	**239738**	$z=a+bx\ln x+cx^2+dx^{2.5}+ey/\ln y+fy^{0.5}$
143	0.9679638395	34	**320522**	$z=a+bx^{1.5}+cx^2+dx^2\ln x+ey^{0.5}\ln y+fy^{0.5}$
144	0.9679638395	20	**133**	$z=a+b\ln x+c(\ln x)^2+d(\ln x)^3+e(\ln x)^4+f/y+g/y^2+h/y^3$
145	0.9679638395	18	**188**	$z=a+b/x+c/x^2+d/x^3+e\ln y+f(\ln y)^2+g(\ln y)^3$
146	0.9679638395	16	**194**	$z=a+b/x+c/x^2+d/x^3+e/y+f/y^2+g/y^3+h/y^4$
147	0.9679638395	12	**192**	$z=a+b/x+c/x^2+d/x^3+e/y+f/y^2$
148	0.9679638395	18	**195**	$z=a+b/x+c/x^2+d/x^3+e/y+f/y^2+g/y^3+h/y^4+i/y^5$
149	0.9679638395	21	**215**	$z=a+b/x+c/x^2+d/x^3+e/x^4+f/x^5+gy+hy^2+iy^3+jy^4+ky^5$
150	0.9679638395	20	**203**	$z=a+b/x+c/x^2+d/x^3+e/x^4+f\ln y+g(\ln y)^2+h(\ln y)^3$
151	0.9679638395	22	**218**	$z=a+b/x+c/x^2+d/x^3+e/x^4+f/x^5+g\ln y+h(\ln y)^2+i(\ln y)^3$
152	0.9679638395	24	**205**	$z=a+b/x+c/x^2+d/x^3+e/x^4+f\ln y+g(\ln y)^2+h(\ln y)^3+i(\ln y)^4+j(\ln y)^5$
153	0.9679638395	26	**220**	$z=a+b/x+c/x^2+d/x^3+e/x^4+f/x^5+g\ln y+h(\ln y)^2+i(\ln y)^3+j(\ln y)^4+k(\ln y)^5$
154	0.9679638395	11	**182**	$z=a+b/x+c/x^2+d/x^3+ey+fy^2$
155	0.9679638395	19	**74**	$z=a+bx+cx^2+dx^3+ex^4+fx^5+g/y+h/y^2+i/y^3+j/y^4$
156	0.9679638395	15	**72**	$z=a+bx+cx^2+dx^3+ex^4+fx^5+g/y+h/y^2$
157	0.9679638395	13	**57**	$z=a+bx+cx^2+dx^3+ex^4+f/y+g/y^2$
158	0.9679638395	24	**128**	$z=a+b\ln x+c(\ln x)^2+d(\ln x)^3+e(\ln x)^4+f\ln y+g(\ln y)^2+h(\ln y)^3$
159	0.9679638395	17	**213**	$z=a+b/x+c/x^2+d/x^3+e/x^4+f/x^5+gy+hy^2+iy^3$
160	0.9679638395	21	**75**	$z=a+bx+cx^2+dx^3+ex^4+fx^5+g/y+h/y^2+i/y^3+j/y^4+k/y^5$
161	0.9679638395	26	**143**	$z=a+b\ln x+c(\ln x)^2+d(\ln x)^3+e(\ln x)^4+f(\ln x)^5+g\ln y+h(\ln y)^2+i(\ln y)^3$
162	0.9679638395	17	**38**	$z=a+bx+cx^2+dx^3+e\ln y+f(\ln y)^2+g(\ln y)^3$
163	0.9679638395	16	**35**	$z=a+bx+cx^2+dx^3+ey+fy^2+gy^3+hy^4+iy^5$

164	0.9679638395	22	**190**	$z=a+b/x+c/x^{2}+d/x^{3}+e\ln y+f(\ln y)^{2}+g(\ln y)^{3}+h(\ln y)^{4}+i(\ln y)^{5}$
165	0.9679638395	15	**198**	$z=a+b/x+c/x^{2}+d/x^{3}+e/x^{4}+fy+gy^{2}+hy^{3}$
166	0.9679638395	22	**225**	$z=a+b/x+c/x^{2}+d/x^{3}+e/x^{4}+f/x^{5}+g/y+h/y^{2}+i/y^{3}+j/y^{4}+k/y^{5}$
167	0.9679638395	16	**222**	$z=a+b/x+c/x^{2}+d/x^{3}+e/x^{4}+f/x^{5}+g/y+h/y^{2}$
168	0.9679638395	20	**224**	$z=a+b/x+c/x^{2}+d/x^{3}+e/x^{4}+f/x^{5}+g/y+h/y^{2}+i/y^{3}+j/y^{4}$
169	0.9679638395	19	**60**	$z=a+bx+cx^{2}+dx^{3}+ex^{4}+f/y+g/y^{2}+h/y^{3}+i/y^{4}+j/y^{5}$
170	0.9679638395	18	**132**	$z=a+b\ln x+c(\ln x)^{2}+d(\ln x)^{3}+e(\ln x)^{4}+f/y+g/y^{2}$
171	0.9679638395	18	**50**	$z=a+bx+cx^{2}+dx^{3}+ex^{4}+fy+gy^{2}+hy^{3}+iy^{4}+jy^{5}$
172	0.9679638395	16	**49**	$z=a+bx+cx^{2}+dx^{3}+ex^{4}+fy+gy^{2}+hy^{3}+iy^{4}$
173	0.9679638395	16	**117**	$z=a+b\ln x+c(\ln x)^{2}+d(\ln x)^{3}+e/y+f/y^{2}$
174	0.9679638395	22	**113**	$z=a+b\ln x+c(\ln x)^{2}+d(\ln x)^{3}+e\ln y+f(\ln y)^{2}+g(\ln y)^{3}$
175	0.9679638395	11	**42**	$z=a+bx+cx^{2}+dx^{3}+e/y+f/y^{2}$
176	0.9679638395	23	**55**	$z=a+bx+cx^{2}+dx^{3}+ex^{4}+f\ln y+g(\ln y)^{2}+h(\ln y)^{3}+i(\ln y)^{4}+j(\ln y)^{5}$
177	0.9679638395	23	**69**	$z=a+bx+cx^{2}+dx^{3}+ex^{4}+fx^{5}+g\ln y+h(\ln y)^{2}+i(\ln y)^{3}+j(\ln y)^{4}$
178	0.9679638395	24	**114**	$z=a+b\ln x+c(\ln x)^{2}+d(\ln x)^{3}+e\ln y+f(\ln y)^{2}+g(\ln y)^{3}+h(\ln y)^{4}$
179	0.9679638395	25	**70**	$z=a+bx+cx^{2}+dx^{3}+ex^{4}+fx^{5}+g\ln y+h(\ln y)^{2}+i(\ln y)^{3}+j(\ln y)^{4}+k(\ln y)^{5}$
180	0.9679638395	20	**119**	$z=a+b\ln x+c(\ln x)^{2}+d(\ln x)^{3}+e/y+f/y^{2}+g/y^{3}+h/y^{4}$
181	0.9679638395	28	**130**	$z=a+b\ln x+c(\ln x)^{2}+d(\ln x)^{3}+e(\ln x)^{4}+f\ln y+g(\ln y)^{2}+h(\ln y)^{3}+i(\ln y)^{4}+j(\ln y)^{5}$
182	0.9679638395	30	**145**	$z=a+b\ln x+c(\ln x)^{2}+d(\ln x)^{3}+e(\ln x)^{4}+f(\ln x)^{5}+g\ln y+h(\ln y)^{2}+i(\ln y)^{3}+j(\ln y)^{4}+k(\ln y)^{5}$
183	0.9679638395	20	**65**	$z=a+bx+cx^{2}+dx^{3}+ex^{4}+fx^{5}+gy+hy^{2}+iy^{3}+jy^{4}+ky^{5}$
184	0.9679638395	28	**144**	$z=a+b\ln x+c(\ln x)^{2}+d(\ln x)^{3}+e(\ln x)^{4}+f(\ln x)^{5}+g\ln y+h(\ln y)^{2}+i(\ln y)^{3}+j(\ln y)^{4}$
185	0.9679638395	23	**139**	$z=a+b\ln x+c(\ln x)^{2}+d(\ln x)^{3}+e(\ln x)^{4}+f(\ln x)^{5}+gy+hy^{2}+iy^{3}+jy^{4}$
186	0.9679638395	14	**207**	$z=a+b/x+c/x^{2}+d/x^{3}+e/x^{4}+f/y+g/y^{2}$
187	0.9679638395	26	**150**	$z=a+b\ln x+c(\ln x)^{2}+d(\ln x)^{3}+e(\ln x)^{4}+f(\ln x)^{5}+g/y+h/y^{2}+i/y^{3}+j/y^{4}+k/y^{5}$
188	0.9679638395	18	**2093**	$z=a+by+cy^{2}+\mathrm{LORCUMX}(d,e,f)$
189	0.9679638395	25	**140**	$z=a+b\ln x+c(\ln x)^{2}+d(\ln x)^{3}+e(\ln x)^{4}+f(\ln x)^{5}+gy+hy^{2}+iy^{3}+jy^{4}+ky^{5}$
190	0.9679638395	10	**32**	$z=a+bx+cx^{2}+dx^{3}+ey+fy^{2}$
191	0.9679638395	12	**47**	$z=a+bx+cx^{2}+dx^{3}+ex^{4}+fy+gy^{2}$
192	0.9679638395	17	**122**	$z=a+b\ln x+c(\ln x)^{2}+d(\ln x)^{3}+e(\ln x)^{4}+fy+gy^{2}$
193	0.9679638395	12	**33**	$z=a+bx+cx^{2}+dx^{3}+ey+fy^{2}+gy^{3}$
194	0.9679638395	15	**107**	$z=a+b\ln x+c(\ln x)^{2}+d(\ln x)^{3}+ey+fy^{2}$
195	0.9679638395	13	**197**	$z=a+b/x+c/x^{2}+d/x^{3}+e/x^{4}+fy+gy^{2}$
196	0.9679638395	16	**187**	$z=a+b/x+c/x^{2}+d/x^{3}+e\ln y+f(\ln y)^{2}$
197	0.9679638395	15	**37**	$z=a+bx+cx^{2}+dx^{3}+e\ln y+f(\ln y)^{2}$
198	0.9679638395	20	**112**	$z=a+b\ln x+c(\ln x)^{2}+d(\ln x)^{3}+e\ln y+f(\ln y)^{2}$
199	0.9679629188	29	**54022**	$z=a+bx^{2}\ln x+cx^{2.5}+d(\ln y)^{2}+ey^{0.5}$
200	0.9679629188	32	**54011**	$z=a+bx^{2}\ln x+cx^{2.5}+dy^{0.5}\ln y+ey^{0.5}$
201	0.9679629188	36	**54010**	$z=a+bx^{2}\ln x+cx^{2.5}+dy^{0.5}\ln y+ey/\ln y$
202	0.9679629188	30	**54032**	$z=a+bx^{2}\ln x+cx^{2.5}+dy/\ln y+ey^{0.5}$
203	0.9679629188	31	**54025**	$z=a+bx^{2}\ln x+cx^{2.5}+d(\ln y)^{2}+e/y^{0.5}$
204	0.9679629188	34	**54026**	$z=a+bx^{2}\ln x+cx^{2.5}+d(\ln y)^{2}+e\ln y/y$
205	0.9679629188	35	**53997**	$z=a+bx^{2}\ln x+cx^{2.5}+de^{y/wy}+e(\ln y)^{2}$
206	0.9679629188	34	**54021**	$z=a+bx^{2}\ln x+cx^{2.5}+d(\ln y)^{2}+ey/\ln y$
207	0.9679629188	30	**54045**	$z=a+bx^{2}\ln x+cx^{2.5}+dy^{0.5}+e\ln y/y$
208	0.9679629188	31	**53983**	$z=a+bx^{2}\ln x+cx^{2.5}+dy^{3}+ey^{0.5}\ln y$
209	0.9679432921	17	**2117**	$z=a+by+cy^{2}+\mathrm{SIGX}(d,e,f)$
210	0.9679220037	55	**2081**	$z=a+by+cy^{2}+\mathrm{GCUMX}(d,e,f)$
211	0.9678975376	22	**2129**	$z=a+by+cy^{2}+\mathrm{EXVCUMX}(d,e,f)$
212	0.9678230376	15	**23**	$z=a+bx+cx^{2}+d\ln y+e(\ln y)^{2}+f(\ln y)^{3}$
213	0.9678230376	9	**27**	$z=a+bx+cx^{2}+d/y+e/y^{2}$
214	0.9678230376	8	**17**	$z=a+bx+cx^{2}+dy+ey^{2}$
215	0.9678230376	13	**22**	$z=a+bx+cx^{2}+d\ln y+e(\ln y)^{2}$
216	0.9677659041	12	**2021**	$z=a+by+cy^{2}+\mathrm{LORX}(d,e,f)$
217	0.9677609336	19	**2165**	$z=a+by+cy^{2}+\mathrm{POWX}(d,e)$
218	0.9677604371	19	**2009**	$z=a+by+cy^{2}+\mathrm{GAUSSX}(d,e,f)$
219	0.9677503846	31	**2045**	$z=a+by+cy^{2}+\mathrm{LOGISTICX}(d,e,f)$
220	0.9677336696	108	**2036**	$z=a+\mathrm{LOGNORMX}(b,c,d)+\mathrm{LOGNORMY}(e,f,g)+\mathrm{LOGNORMX}(h,c,d)*\mathrm{LOGNORMY}(1,f,g)$

221	0.9677207969	27	**2057**	$z=a+by+cy^2+EXTRVALX(d,e,f)$
222	0.9673253305	13	**2014**	$z=a+LORX(b,c,d)*LORY(1,e,f)$
223	0.9672704333	44	**2038**	$z=a+LOGISTICX(b,c,d)*LOGISTICY(1,e,f)$
224	0.9672250305	61	**2105**	$z=a+by+cy^2+LNCUMX(d,e,f)$
225	0.9671164876	19	**6900**	$z=a+bx^2+c/y^{1.5}+dlny/y^2$
226	0.9671164876	21	**6879**	$z=a+bx^2+c/lny+dlny/y$
227	0.9671164876	16	**6841**	$z=a+bx^2+c(lny)^2+dy^{0.5}$
228	0.9671164876	18	**6830**	$z=a+bx^2+cy^{0.5}lny+dy^{0.5}$
229	0.9671164876	23	**6829**	$z=a+bx^2+cy^{0.5}lny+dy/lny$
230	0.9671164876	17	**6851**	$z=a+bx^2+cy/lny+dy^{0.5}$
231	0.9671164876	18	**6903**	$z=a+bx^2+clny/y^2+d/y^2$
232	0.9671164876	21	**6843**	$z=a+bx^2+c(lny)^2+d/lny$
233	0.9671164876	18	**6844**	$z=a+bx^2+c(lny)^2+d/y^{0.5}$
234	0.9671164876	21	**6845**	$z=a+bx^2+c(lny)^2+dlny/y$
235	0.9670894619	21	**2002**	$z=a+GAUSSX(b,c,d)*GAUSSY(1,e,f)$
236	0.9669576438	41	**2063**	$z=a+GAUSSX(b,c,d)*LOGNORMY(1,e,f)$
237	0.9666677741	25	**2090**	$z=a+LORCUMX(b,c,d)+LORCUMY(e,f,g)$
238	0.9666610231	26	**2114**	$z=a+SIGX(b,c,d)+SIGY(e,f,g)$
239	0.9666121789	102	**2078**	$z=a+GCUMX(b,c,d)+GCUMY(e,f,g)$
240	0.9665583973	37	**2126**	$z=a+EXVCUMX(b,c,d)+EXVCUMY(e,f,g)$
241	0.9665236106	32	**2024**	$z=a+LORX(b,c,d)+LORY(e,f,g)+LORX(h,c,d)*LORY(1,f,g)$
242	0.9663810768	55	**2082**	$z=a+bx+cx^2+GCUMY(d,e,f)$
243	0.966378426	23	**2142**	$z=a+bx+cx^2+LDRY(d,e,f)$
244	0.9663727498	18	**2094**	$z=a+bx+cx^2+LORCUMY(d,e,f)$
245	0.9662862378	38	**2138**	$z=a+LDRX(b,c,d)+LDRY(e,f,g)$
246	0.966266393	22	**2130**	$z=a+bx+cx^2+EXVCUMY(d,e,f)$
247	0.9657625458	37	**2137**	$z=LDRX(a,b,c)+LDRY(d,e,f)$
248	0.965676567	114	**2101**	$z=LNCUMX(a,b,c)+LNCUMY(d,e,f)$
249	0.9656121531	115	**2102**	$z=a+LNCUMX(b,c,d)+LNCUMY(e,f,g)$
250	0.963953427	101	**2077**	$z=GCUMX(a,b,c)+GCUMY(d,e,f)$
251	0.9635868418	14	**102**	$z=a+blnx+c(lnx)^2+d/y+e/y^2$
252	0.9635868418	13	**92**	$z=a+blnx+c(lnx)^2+dy+ey^2$
253	0.9635868418	18	**97**	$z=a+blnx+c(lnx)^2+dlny+e(lny)^2$
254	0.9630989621	50	**2096**	$z=a+LORCUMX(b,c,d)+LORCUMY(e,f,g)+LORCUMX(h,c,d)*LORCUMY(1,f,g)$
255	0.9623727334	25	**2113**	$z=SIGX(a,b,c)+SIGY(d,e,f)$
256	0.96187202	19	**2010**	$z=a+bx+cx^2+GAUSSY(d,e,f)$
257	0.9618674235	30	**2006**	$z=a+GAUSSX(b,c,d)+GAUSSY(e,f,g)$
258	0.9618648939	53	**2042**	$z=a+LOGISTICX(b,c,d)+LOGISTICY(e,f,g)$
259	0.9618599703	16	**2018**	$z=a+LORX(b,c,d)+LORY(e,f,g)$
260	0.9618596392	12	**2022**	$z=a+bx+cx^2+LORY(d,e,f)$
261	0.9618061236	31	**2046**	$z=a+bx+cx^2+LOGISTICY(d,e,f)$
262	0.9594527163	54	**2030**	$z=a+LOGNORMX(b,c,d)+LOGNORMY(e,f,g)$
263	0.9594272849	42	**2068**	$z=a+LOGNORMX(b,c,d)+GAUSSY(e,f,g)$
264	0.9506058759	84	**2072**	$z=a+LOGNORMX(b,c,d)+GAUSSY(e,f,g)+LOGNORMX(h,c,d)*GAUSSY(1,f,g)$
265	0.948986761	36	**2125**	$z=EXVCUMX(a,b,c)+EXVCUMY(d,e,f)$
266	0.9485691185	13	**8**	$z=a+bx+clny+d(lny)^2+e(lny)^3$
267	0.9485691185	9	**13**	$z=a+bx+c/y+d/y^2+e/y^3$
268	0.9485691185	7	**12**	$z=a+bx+c/y+d/y^2$
269	0.9485691185	8	**3**	$z=a+bx+cy+dy^2+ey^3$
270	0.9485691185	6	**2**	$z=a+bx+cy+dy^2$
271	0.9485691185	11	**7**	$z=a+bx+clny+d(lny)^2$
272	0.9485690615	15	**2153**	$z=a+by+cy^2+EXPX(d,e)$
273	0.9476107325	17	**2008**	$z=a+bx+GAUSSY(c,d,e)$
274	0.9476098295	25	**2056**	$z=a+bx+EXTRVALY(c,d,e)$
275	0.9476089329	10	**2020**	$z=a+bx+LORY(c,d,e)$
276	0.9476080688	29	**2044**	$z=a+bx+LOGISTICY(c,d,e)$
277	0.9476042135	29	**2032**	$z=a+bx+LOGNORMY(c,d,e)$
278	0.9465218201	10	**177**	$z=a+b/x+c/x^2+d/y+e/y^2$
279	0.9455359516	34	**501**	Fourier Series Simple Order 2x1
280	0.9401527484	20	**2128**	$z=a+bx+EXVCUMY(c,d,e)$

281	0.9401527484	53	**2080**	z=a+bx+GCUMY(c,d,e)
282	0.9401527336	14	**2092**	z=a+bx+LORCUMY(c,d,e)
283	0.9401487612	21	**2140**	z=a+bx+LDRY(c,d,e)
284	0.9401474415	59	**2104**	z=a+bx+LNCUMY(c,d,e)
285	0.9401418053	15	**2116**	z=a+bx+SIGY(c,d,e)
286	0.9371466254	7	**3144**	$z=a+bx^2+cy^3$
287	0.9330049004	11	**3217**	$z=a+bx^{2.5}+cy^3$
288	0.9313697538	10	**3143**	$z=a+bx^2+cy^{2.5}$
289	0.9293730395	13	**3067**	$z=a+bx\ln x+cy^3$
290	0.9283355588	13	**3142**	$z=a+bx^2+cy^2\ln y$
291	0.9278470564	10	**3106**	$z=a+bx^{1.5}+cy^3$
292	0.9272409695	13	**3216**	$z=a+bx^{2.5}+cy^{2.5}$
293	0.9270606539	14	**3181**	$z=a+bx^2\ln x+cy^3$
294	0.92586455	60	**2168**	z=a+POWX(b,c)+POWY(d,e)+POWX(f,c)*POWY(1,e)
295	0.9257428396	15	**3286**	$z=a+be^{x/wx}+cy^3$
296	0.9249264002	6	**3141**	$z=a+bx^2+cy^2$
297	0.9239143001	15	**151232650**	$\ln z=a+bx^{0.5}+cy^3$
298	0.9195126968	22	**151232557**	$\ln z=a+bx^{0.5}\ln x+cy^3$
299	0.9189204163	19	**151232679**	$\ln z=a+b\ln x+cy^3$
300	0.9171007228	18	**151232649**	$\ln z=a+bx^{0.5}+cy^{2.5}$
301	0.9160159203	13	**151232265**	$\ln z=a+bx+cy^3$
302	0.9138326689	21	**151232648**	$\ln z=a+bx^{0.5}+cy^2\ln y$
303	0.9133451901	30	**2162**	z=a+POWX(b,c)+POWY(d,e)
304	0.9127405688	14	**2091**	z=a+by+LORCUMX(c,d,e)
305	0.912661092	105	**2047**	z=LOGISTICX(a,b,c)+LOGISTICY(d,e,f)+LOGISTICX(g,b,c)*LOGISTICY(1,e,f)
306	0.9126052013	15	**2115**	z=a+by+SIGX(c,d,e)
307	0.9124671815	24	**151232556**	$\ln z=a+bx^{0.5}\ln x+cy^{2.5}$
308	0.912463175	53	**2079**	z=a+by+GCUMX(c,d,e)
309	0.9124008415	21	**151232678**	$\ln z=a+b\ln x+cy^{2.5}$
310	0.9123873469	20	**2127**	z=a+by+EXVCUMX(c,d,e)
311	0.9119597912	17	**2163**	z=a+by+POWX(c,d)
312	0.9119543489	6	**16**	$z=a+bx+cx^2+dy$
313	0.9119225976	21	**2139**	z=a+by+LDRX(c,d,e)
314	0.9118448597	29	**2043**	z=a+by+LOGISTICX(c,d,e)
315	0.9118285414	10	**2019**	z=a+by+LORX(c,d,e)
316	0.9117911512	17	**2007**	z=a+by+GAUSSX(c,d,e)
317	0.9117863162	52	**2041**	z=LOGISTICX(a,b,c)+LOGISTICY(d,e,f)
318	0.9117191415	25	**2055**	z=a+by+EXTRVALX(c,d,e)
319	0.9110734175	21	**2150**	z=a+EXPX(b,c)+EXPY(d,e)
320	0.9110032565	59	**2103**	z=a+by+LNCUMX(c,d,e)
321	0.9103750006	14	**151232647**	$\ln z=a+bx^{0.5}+cy^2$
322	0.910042532	25	**2086**	z=a+LORCUMX(b,c,d)*LORCUMY(1,e,f)
323	0.9088625816	15	**151232264**	$\ln z=a+bx+cy^{2.5}$
324	0.9078479293	59	**2167**	z=POWX(a,b)+POWY(c,d)+POWX(e,b)*POWY(1,d)
325	0.9076461218	21	**2146**	z=a+EXPX(b,c)*EXPY(1,d)
326	0.9063727753	21	**2004**	z=a+GAUSSX(b,c,d)*GAUSSY(1,e,d)
327	0.9051435989	101	**2073**	z=GCUMX(a,b,c)*GCUMY(1,d,e)
328	0.9047033886	25	**2109**	z=SIGX(a,b,c)*SIGY(1,d,e)
329	0.9041917779	11	**91**	$z=a+b\ln x+c(\ln x)^2+dy$
330	0.903916449	9	**302462023**	$z^{-1}=a+b/x+cy^3$
331	0.9028758051	11	**21**	$z=a+bx+cx^2+d\ln y$
332	0.9023180814	12	**302462047**	$z^{-1}=a+b/x^{1.5}+cy^3$
333	0.9009931162	15	**302462113**	$z^{-1}=a+be^{-x}+cy^3$
334	0.9007742856	27	**2058**	$z=a+bx+cx^2$+EXTRVALY(d,e,f)
335	0.9000474283	30	**2158**	z=a+POWX(b,c)*POWY(1,d)
336	0.9000333898	44	**2040**	z=a+LOGISTICX(b,c,d)*LOGISTICY(1,e,d)
337	0.8989221327	7	**26**	$z=a+bx+cx^2+d/y$
338	0.8988211735	44	**2054**	z=a+EXTRVALX(b,c,d)+EXTRVALY(e,f,g)
339	0.898103356	29	**2157**	z=POWX(a,b)*POWY(1,c)
340	0.8980967113	24	**2085**	z=LORCUMX(a,b,c)*LORCUMY(1,d,e)
341	0.8977069711	61	**2106**	$z=a+bx+cx^2$+LNCUMY(d,e,f)

342	0.8974371898	12	**302462022**	z^{-1}=a+b/x+$cy^{2.5}$
343	0.8969027168	44	**2052**	z=a+EXTRVALX(b,c,d)*EXTRVALY(1,e,d)
344	0.8969027157	44	**2050**	z=a+EXTRVALX(b,c,d)*EXTRVALY(1,e,f)
345	0.8966345111	16	**96**	z=a+blnx+c$(\ln x)^2$+dlny
346	0.8963056349	29	**2005**	z=GAUSSX(a,b,c)+GAUSSY(d,e,f)
347	0.8961084131	29	**2161**	z=POWX(a,b)+POWY(c,d)
348	0.8959681977	14	**302462046**	z^{-1}=a+b/$x^{1.5}$+$cy^{2.5}$
349	0.8954144194	24	**2089**	z=LORCUMX(a,b,c)+LORCUMY(d,e,f)
350	0.8946497926	17	**302462112**	z^{-1}=a+be^{-x}+$cy^{2.5}$
351	0.8942903588	15	**302462021**	z^{-1}=a+b/x+cy^2lny
352	0.8937978571	12	**101**	z=a+blnx+c$(\ln x)^2$+d/y
353	0.8932371316	12	**302461972**	z^{-1}=a+b/$x^{0.5}$+cy^3
354	0.893044633	37	**2133**	z=LDRX(a,b,c)*LDRY(1,d,e)
355	0.8929021029	17	**302462045**	z^{-1}=a+b/$x^{1.5}$+cy^2lny
356	0.8915889966	20	**302462111**	z^{-1}=a+be^{-x}+cy^2lny
357	0.8911562467	114	**2099**	z=LNCUMX(a,b,c)*LNCUMY(1,d,c)
358	0.8909535679	17	**2118**	z=a+bx+cx^2+SIGY(d,e,f)
359	0.8894458858	37	**2135**	z=LDRX(a,b,c)*LDRY(1,d,c)
360	0.8890919137	31	**2034**	z=a+bx+cx^2+LOGNORMY(d,e,f)
361	0.8865694305	42	**2067**	z=a+GAUSSX(b,c,d)+LOGNORMY(e,f,g)
362	0.8858171729	13	**2016**	z=a+LORX(b,c,d)*LORY(1,e,d)
363	0.8818371752	13	**2151**	z=a+by+EXPX(c,d)
364	0.8818371135	4	**2169**	z=a+bx+cy [Robust None, Least Squares]
365	0.8818371135	4	**1**	z=a+bx+cy
366	0.8799604124	7	**166**	z=a+b/x+c/x^2+dy
367	0.8785071761	4	**2170**	z=a+bx+cy [Robust Low, Least Abs Deviation]
368	0.8779277927	13	**2152**	z=a+bx+EXPY(c,d)
369	0.8778961102	17	**2164**	z=a+bx+POWY(c,d)
370	0.8775423246	9	**6**	z=a+bx+clny
371	0.8769941665	5	**11**	z=a+bx+c/y
372	0.8697730427	4	**2171**	z=a+bx+cy [Robust Medium, Lorentzian]
373	0.8695566757	4	**2172**	z=a+bx+cy [Robust High, PearsonVII Limit]
374	0.8670609827	15	**2154**	z=a+bx+cx^2+EXPY(d,e)
375	0.8282892003	11	**1049**	z=(a+bx+cx^2+dy)/(1+ex+fy)
376	0.8241383598	19	**2166**	z=a+bx+cx^2+POWY(d,e)
377	0.7997599516	10	**86**	z=a+blnx+c/y
378	0.7952226016	14	**81**	z=a+blnx+clny
379	0.7927837275	14	**1129**	z=(a+blnx+cy)/(1+dlnx+ey)
380	0.7920628183	9	**76**	z=a+blnx+cy
381	0.7810052147	19	**1153**	z=(a+blnx+clny)/(1+dlnx+elny)
382	0.6990402781	6	**161**	z=a+b/x+c/y
383	0.6896166317	10	**156**	z=a+b/x+clny
384	0.6787310167	5	**151**	z=a+b/x+cy
385	0.6663407439	14	**1025**	z=(a+bx+clny)/(1+dx+elny)
386	0.6576159725	43	**2037**	z=LOGISTICX(a,b,c)*LOGISTICY(1,d,e)
387	0.6340947854	38	**2134**	z=a+LDRX(b,c,d)*LDRY(1,e,f)
388	0.6340837244	115	**2100**	z=a+LNCUMX(b,c,d)*LNCUMY(1,e,d)
389	0.6340824897	115	**2098**	z=a+LNCUMX(b,c,d)*LNCUMY(1,e,f)
390	0.633807525	36	**2122**	z=a+EXVCUMX(b,c,d)*EXVCUMY(1,e,f)
391	0.6335318381	38	**2136**	z=a+LDRX(b,c,d)*LDRY(1,e,d)
392	0.6330075579	102	**2076**	z=a+GCUMX(b,c,d)*GCUMY(1,e,d)
393	0.6315232298	26	**2112**	z=a+SIGX(b,c,d)*SIGY(1,e,d)
394	0.6284578258	25	**2088**	z=a+LORCUMX(b,c,d)*LORCUMY(1,e,d)
395	0.6237117244	102	**2074**	z=a+GCUMX(b,c,d)*GCUMY(1,e,f)
396	0.569708431	45	**2026**	z=a+LOGNORMX(b,c,d)*LOGNORMY(1,e,f)
397	0.4253421737	24	**2087**	z=LORCUMX(a,b,c)*LORCUMY(1,d,c)
398	0.4192154759	26	**2110**	z=a+SIGX(b,c,d)*SIGY(1,e,f)
399	0.4187905231	36	**2121**	z=EXVCUMX(a,b,c)*EXVCUMY(1,d,e)
400	0.4187905231	37	**2123**	z=EXVCUMX(a,b,c)*EXVCUMY(1,d,c)
401	0.4185794791	101	**2075**	z=GCUMX(a,b,c)*GCUMY(1,d,c)
402	0.4180084961	25	**2111**	z=SIGX(a,b,c)*SIGY(1,d,c)
403	0.4127058533	114	**2097**	z=LNCUMX(a,b,c)*LNCUMY(1,d,e)

404	0.3751489737	31	**2023**	z=LORX(a,b,c)+LORY(d,e,f)+LORX(g,b,c)*LORY(1,e,f)
405	0.3557023573	59	**2011**	z=GAUSSX(a,b,c)+GAUSSY(d,e,f)+GAUSSX(g,b,c)*GAUSSY(1,e,f)
406	0.3318013525	23	**2141**	z=a+by+cy^2+LDRX(d,e,f)
407	0.3271618105	107	**2035**	z=LOGNORMX(a,b,c)+LOGNORMY(d,e,f)+LOGNORMX(g,b,c)*LOGNORMY(1,e,f)
408	0.1870071253	41	**2064**	z=a+LOGNORMX(b,c,d)*GAUSSY(1,e,f)

INDEX

Note: Page numbers followed by "*f*" and "*t*" refer to figures and tables, respectively.

A

B

C

D

E

P

Q

R

T

U

V